ENCYCLOPAEDIA OF HORTICULTURAL CROPS

Volume-1

ORGANIC VEGETABLE PRODUCTION

S. K. GUPTA

RAJAT PUBLICATIONS
NEW DELHI - 110 002 (INDIA)

RAJAT PUBLICATIONS
4740/23, Ansari Road, Daryaganj
New Delhi- 110 002 (India)
Phones : +91-11-23267924, +91-9811484973
E-mail: rajat_publication@yahoo.com

Encyclopaedia of Horticultural Crops

Edition 2025

ISBN 978-81-7880-319-7 (Set)

[The responsibility for facts stated opinion expressed or conclusions reached and plagiarism, if any, in this volume is entirely that of the Editor. The publisher bears no responsibility for them whatsoever.]

PRINTED IN INDIA

Published by Mrs. Seema Wasan for Rajat Publications, New Delhi and Printed at H.S. Offset Press, Delhi.

Preface

Horticulture refers to the area of agricultural science that deals with garden plant cultivation, the science and out of growing fruits, vegetables, ornamental trees bushes and flowers, on a commercial scale, in nurseries, orchards and gardens. Along with forestry and agronomy, horticulture constitutes the three broad branches of agricultural science and practice. As importance and relevance of aggressive cannot be ignored, so can that of horticulture be also not ignored. In fact, with the increase in demand for newer and better varieties of horticultural products, the significance of horticulture has increased all the more.

This encyclopedia of horticultural crops is a three volume set which brings a compendious compilation of information on the various forms science and practice of cultivating horticultural crops. Through these three works, foundation has been laid by the encyclopedia for a discussion of the principles of horticultural science, what it comprises of, its history and its present, and what the future holds for this field.

The encyclopedia commences with the book organic vegetable production a book which seeks to discuss the cultivation of organic vegetables in agro-science. As organic farming and its products have entered the market in a significant manner, it becomes imperative that a manual seek the principles of this practice and therein lies the purpose of this book. Cultivation of Herbs and Flowers as the title suggests, is a work bringing forth a comprehensive detailing of cultivating herbs and flower which after vegetables, consultant a major part of the demand for horticultural products, both for purposes in practical applications (as in medicinal), and for ornamentation. The concluding work in this series Is the book Horticultural Crop Production. As many horticultural crops today particularly vegetables—are often grown as field crops and considered agricultural commodities, it has become increasingly evident that horticulture products are much more importance than as mere garden products. This book seeks a study of horticultural science as a whole, with special emphasis on its products as "Crops".

All these three books aim at providing a reliable encylopaedia on horticultural productions; and provide material for through study of the science itself. It is hoped that the comprehensive and informative content of this series foster an enriching reading and learning experience for readers.

S.K. Gupta

Contents

1

Organic Farming

The International Federation of Organic Agriculture Movements (IFOAM), a private body that sets standards for organic agriculture, has 750 member organisations in more than 100 countries. Codex Alimentarius Commission, the United Nations body that oversees the world's food standards, deliberated long and hard to define organic agriculture as a "holistic production management system that avoids use of synthetic fertilisers and pesticides, minimizes pollution of air, soil and water, and optimizes the health and productivity of independent communities of life, plants, animals and people".

Organic agriculture has its roots in traditional agricultural practices that evolved in countless villages and farming communities over the millennia. By trial-and-error local farmers passed their best results from generation to generation. The modern face of organic agriculture emerged in the late 1960s, when farmers and consumers began to recognise that the enormous amount of chemicals being used in both crop and animal production could have dire consequences for the earth and its people. Although its initial modern image was of health nuts and hippies, organic agriculture has moved into the mainstream. It is now the fastest growing food sector. From coffee in Mexico to tea in China, from shrimp in Thailand to beef in Argentina, from cotton in Uganda to olives in Italy—today's organic farmers build on the past, using modern scientific research to guide them in upgrading their traditional farming methods.

Science can now explain how and why those methods worked, opening avenues for even further improvement. New information emerges constantly from scientific research, uncovering nature's incredible secrets like pieces in a gigantic ecological puzzle · bees pass through natural foot baths in their hives that reduce mould in strawberries; bats, long considered crop pests, actually provide many services such as WHAT IS IT? organic agriculture: pollinating, dispersing seeds and controlling insects; earthworms, bacteria and fungi serve as soil "engineers" and create conditions for water and nutrients to circulate. The more farmers know about these puzzle pieces, the better their ability to manage the natural elements and benefit.

The Organic Movement Exists Amidst Controversy. Doubters express concerns that without synthetic fertilisers and pesticides there will not be enough yields to feed the earth's growing population, that the market is driven by developed countries whose strict regulations cannot be met by developing-country producers, that organic agriculture's labour-intensive production methods increase drudgery.

Movement supporters point out that crop yields often decline after extended chemical use, increased labour requirements of organic agriculture balance with decreased expenditures for inputs, organic products do not have to be certified to be legitimate, and that there is no way to put a monetary value on conserving the earth's biodiversity and fragile ecosystems as well as its local cultures and culinary traditions.

Absence of synthetic inputs is not the only prerequisite of organic agriculture. In fact, traditional systems that are "negligent", that carelessly mine soil nutrients or overgrase pastures, would not qualify for any organic status, even if there are no inputs present. Traditional farms do not need to be certified to be considered "organic"; when farmers follow organic management principles, thus producing food while conserving natural resources, they qualify as "non-certified organic".

The type of production methods organic farmers choose depends on more than agro-environmental conditions. It also depends on socioeconomic situations, such as their labour and investment capacity and on the targeted market. Organic farmers put great efforts into study and training, seeking out developments appropriate for their farms, adapting them, and eventually going through inspection and certification to access markets, all aimed at creating a sustainable system throughout the food supply chain. The goal is for fields and animals to produce to their maximum and for farmers to protect the quality of their air, soil and streams.

Organic Agriculture and Community Development

The organic agriculture movement offers farmers new possibilities for helping themselves. Although considered modern, organic agriculture actually is based on traditional agriculture knowledge and culture. Over the centuries, traditional agriculture developed because small farmers experimented, innovated, adapted and worked together. They shared their seeds and new ideas about how to improve production. Today's organic agriculture has the same evolution, recognising there is no one way to address a problem, every case has a specific response, and the concerned farmer is the one who knows best. Farming methods evolve to match local environments, responding to unique biophysical and socioeconomic constraints and opportunities. Increased trade, with its exchange of goods, capital and labour, has also meant the erosion of small farmers' knowledge and collapse of rural institutions. Too often, decisions on what to grow or how to farm have been decided by outside "experts".

Organic agriculture, with its emphasis on local resources and local ecological knowledge, brings farmers together in their communities. The market demand for organic products, as well as the opportunity to develop agri-tourism for city dwellers who appreciate a healthy, diversified rural landscape, have created new income opportunities for organic farmers.

Organic farmers' and consumers' groups work to support markets, cut out monopolies and increase farm incomes. Organic farmers, both in developed and developing nations, recognise the need to join with their neighbours, set up support networks, share what they know, pool their resources, establish themselves as an organised force and, of course, provide quality products.

Consumers increasingly demand both environmentally friendly and socially just products which means that organic certification also requires proof of fair wages, healthy working conditions and the workers' right of association. Organic boxes manufactured by disabled persons, Canada With the organic movement, it is more important than ever for farmers to be connected. Organic farmers cannot operate in isolation because in order to maintain the integrity of their air, water and soil, they depend on their neighbours' farming practices.

The deep roots of the organic agriculture movement connect farmers, consumers and their markets, improving economic conditions and creating a vibrant rural community.

FOOD SECURITY

Most farmers in developing countries are poor. They either cannot afford or cannot find the necessary inputs to improve their production, and they are not connected to markets to buy or sell food. So as the large landholders who have the resources to purchase the inputs and learn to use them are able to increase their holdings and their incomes, and as international trade brings down prices even in local markets, small farmers become less and less able to compete or, in some cases, even to survive.

Ironically, one solution for these small farmers has been on their farms for centuries. What is now called "organic agriculture", put forth as a modern solution to environmental problems caused by chemical-dependent farming, is actually an improved form of the farming traditionally practiced by small farmers—farming that relies on efficient use of available resources such as soil, water, local breeds and varieties of animals and crops, and a mix of modern and traditional ecological knowledge.

For countless generations, small farmers in developing countries have inherited and managed complex farming systems adapted to their local conditions. These individual farming systems have helped them survive harsh environments, meet their family's food needs, contribute to local/regional markets and still conserve their natural resources. Many of these farmers would be amazed to know that what they learned from their parents is now part of a modern, global movement.

Another irony is that many small farms that never had the advantage of technological advances could now be considered "organic", although "organic by default". An estimated 450 million farms were bypassed by Green Revolution technologies such as mechanisation, irrigation, improved seeds and breeds, and synthetic fertilisers and pesticides.

These include countless low-level technology systems which feature polycultures, agroforestry, integrated crop-livestock systems and productive agro-ecosystems well adapted to local conditions. The absence of these technologies does not automatically make farms organic, but makes it easier for interested farmers to upgrade their farms to organic by adding modern scientific organic practices to their traditional farming methods.

Organic agriculture stresses diversification and adaptive management which significantly decreases vulnerability to weather vagaries or other factors. In spite of a misconception that agro-ecological systems cannot increase agricultural productivity, the reality is that multicropping increases yields significantly. A diversified organic farming system increases farm production by 20 to 60 percent as compared to a traditional low-input system and, at the same time, improves stability by improving soil and water quality and the ecological services that support agriculture.

These methods may not achieve the same yields as the high input systems, but they provide a path to increase both yields and incomes. Farmers can use available resources without fear of major losses, such as those incurred by farmers who often see their expensive inputs washed away by unexpected rains.

It all adds up to a direction for improving food security that is highly promising. Farmers do not need to be certified to know that if they adopt organic agricultural systems and yields become healthier, more dependable and plentiful, their ability to feed their families, either with the food they produce or income from products they sell, will also increase.

Even in areas with poor resources and no access to markets, farmers can improve their family food supply if they have access to extension or training to re-learn to farm organically, and are willing to replace the external inputs with their own labour. The result will be an improved return on their labour investment in terms of better income.

Creating a Nutrient Cycle

Forty percent of the earth's land is agricultural cropland, managed forests or pasture, so it stands to reason that if agriculture does not exist in harmony with the rest of the earth's environment, the potential for widespread environmental degradation caused by loss of habitats and forests, and by pollution of air, soil and water from agricultural chemicals, looms quite large.

This enormous piece of the earth's real estate both affects, and is affected by, climate change and quality of soil, water and air. Inappropriate agricultural practices account for 28 percent of the soil erosion of the past 50 years. Scientists warn that as global warming becomes more and more of a recognisable reality, cultivation zones will shift, and pests and diseases will mutate and proliferate. Through intensification of livestock systems, nitrogen fertilisation and irrigation, agriculture contributes to more than 20 percent of global anthropogenic greenhouse gas emissions.

Soil fertility is the cornerstone of organic management: Because organic farmers do not use synthetic nutrients to restore degraded soil, they must concentrate on building and maintaining soil fertility primarily through their basic farming practices. They depend on multicropping systems and crop rotations, cover crops, organic fertilisers and minimum tillage to maintain and improve soil quality.

The natural fertilisers they use, such as green manure, farmyard manure, compost and plant residues, build organic content and increase the soil's capacity to circulate nutrients, air and water. As crops use soil nutrients, they can be replaced with natural rock minerals such as potassium, phosphate, calcium, magnesium and other trace elements from external sources. Organic agriculture stresses careful management to meet crop needs and avoid excess application of manure and other organic matter that could cause nitrate leaching.

In Europe, soils managed organically have 30 to 40 percent more biomass and 30 to 100 percent more microbial activity than soils managed conventionally. In some developed countries, government waterworks encourage conversion to organic agriculture to reduce the cost of purifying drinking water. In many developing countries, organically managed soils have substantially less erosion and better moisture holding capacity—an essential factor in rain-fed agriculture.

In organic agriculture, the restricted use of mineral fertilisers reduces the use of non-renewable energy (fossil fuels) and reduces the emissions of agricultural greenhouse gases (carbon dioxide, nitrous dioxide and methane). Moreover, mixed farming and soil building allow for increased biological activity by providing support for microorganisms, earthworms, fungi and bacteria. Soils enriched with fauna and flora not only increase nutrient cycling and agricultural productivity but stabilise soils against erosion and floods, detoxify ecosystems and may even help counteract climate change by restoring "soil's capacity for carbon sequestration".

DESERTIFICATION

Desertification, or the degradation of drylands, is caused by overcultivation, overgrasing and deforestation, and results in soil exhaustion and soil erosion. It diminishes soil productivity, reduces food production and deprives land of its vegetative cover. It also has negative affects on areas not directly involved, because desertification can also lead to floods, soil salinisation, deterioration of water quality, and silting of rivers, streams and reservoirs.

Organically managed soils are more resilient both to water stress and nutrient loss. Because of this, they have the potential to counter soil degradation. Organic farmers feed their fields with organic matter that enhances degraded soils. Microorganisms have a good feeding base and create a stable soil structure. Water and nutrient retention capacity increases thanks to a high level of organic matter and permanent soil cover, allowing a substantial reduction in the amount of water needed for irrigation.

Although there has been little specific research on organic agriculture's potential for combating desertification, several practical trials of organic agriculture systems in arid areas have found that organic agriculture helps bring degraded lands back to fertility and create suitable microclimates in dry areas. In organic agriculture, permanent soil cover is an intrinsic part the system. Locally adapted leguminous crops have been found to restore degraded soils very quickly, suppress weeds, fix nitrogen and prevent erosion. When properly set up and adapted to soil and climatic conditions, they do not compete with the main crops for nutrients or water.

Even though the adoption of organic agriculture seems a viable alternative in arid areas, there is a range of constraints to adopting organic agriculture systems or even individual techniques. These include a lack of knowledge, scarce availability of organic materials, insecure land tenure, the perception of organic agriculture as being old fashioned, and the fact that extension services promote a way of farming dependent on chemical inputs.

Biodiversity

The organic agriculture movement, with its focus on quality, nutrition, resistance to stress factors such as poor soils, harsh climates or diseases, its reliance on locally adapted plants and animals, and its goal of increasing yields, stands as an oasis in the midst of that world. Organic farmers must commit to establish a minimum level of biodiversity to compensate for the restriction on synthetic input usage.

Valuing and improving farmers' indigenous knowledge is a very important part of today's organic agriculture movement. Traditional farming, fishing and forest communities not only understand and can apply good ecological theory in decisionmaking, they understand global environmental threats and their potential impact on their livelihoods. Farmers are not trained scientists, but they are ecologically literate. Their knowledge has been acquired through on-farm experimenting, observation of natural processes, and sharing.

Throughout the centuries, farmers have developed local breeds and species of plants and animals, nurtured them and adapted them to local conditions. They saved their own seeds for planting from year to year. When their crops were inundated by pests or there were harsh weather conditions, they saved the seeds of the most resistant plants so that their resistance would be manifest in future generations.

In difficult climatic conditions and in the absence of veterinary services, farmers bred local animals to survive adversity and resist diseases. The primary production strategy of organic farmers is to maintain traditional plants and animals that may yield less in the short run but will be more resilient and better able to survive in the long run.

Diversified organic farming systems have built-in insurance policies—if one crop fails, others will provide sufficient food for the family. Wild species also provide services within organic systems such as pollination, pest control and maintenance of soil fertility, all of which increase biodiversity and, in turn, support agricultural production.

Consumer demand for traditional and speciality products creates new market opportunities and ensures the economic viability of traditional products. For example, the commercial revival of organic quinoa and naturally coloured cotton from Peru, of organic chocolate processed according to an ancient Mayan recipe in Mexico, of local rice varieties in Indonesia, and of native poultry in South Africa has created connections between marginalised producers and the market. Organic agriculture's economic valorisation of endangered genetic resources is key to the survival of poor rural communities in marginal areas as well as of under-utilised species and varieties.

Validity of the Organic System

With organic products, the label tells a big story. Informed consumers know that the label will tell them everything from the name and type of organisation that certified the product's "organic" validity to the type of standards that the product met throughout its growth and processing. These standards make organic agriculture the food industry's most well defined food supply system.

Certification bodies ascertain that products meet the standards set by either private or public institutions. In countries that have organic legislation, certification bodies are approved and supervised by the government. Some also are accredited by the International Organic Accreditation Service (IOAS), an independent non-governmental organisation that authorises certifiers who follow the voluntary standards of the International Federation of Organic Agriculture Movements (IFOAM).

Administered jointly by the Food and Agriculture Organisation of the United Nations (FAO) and World Health Organisation (WHO), the Codex Alimentarius Commission develops and adopts international food standards, guidelines and related texts that serve as the basis for many national standards. It governs such crucial issues as food labelling, residues of pesticides and veterinary drugs, food hygiene, food import and export inspection and certification systems.

In 1999, Codex adopted guidelines on the Production, processing, labelling and marketing of organically produced foods that provide a description of the "organic" system and the "organic" labelling claim in order to ensure fair trade and to facilitate the development of the organic sector. The private sector equivalent of the Codex Alimentarius guidelines is the International basic standards for organic production and processing, created by IFOAM. Both the Codex and IFOAM guidelines cover the production of plants, livestock and bees; the handling, storage, processing, packaging and transportation of products; and a list of substances permitted in the production and processing of organic foods.

IFOAM also has provisions for fibres, aquaculture and non-wood forest products. protect consumers against deception and fraud; protect organic products from misrepresentation by non-organic products; ensure inspection and compliance throughout the process of production, processing, storage, transport and marketing; harmonise provisions for the production, certification and labelling of organic produce; facilitate the establishment of equivalence agreements between importing and exporting countries; and enhance organic agricultural systems locally and globally.

The Codex and IFOAM guidelines for organic agriculture are minimal, intended to guide governments and private certification bodies in setting standards. In other words, they are "standards for standards" meant to:

> protect consumers against deception and fraud; protect organic products from misrepresentation by non-organic products; ensure inspection and compliance throughout the process of production, processing, storage, transport and marketing; harmonise provisions for the production, certification and labelling of organic produce; facilitate the establishment of equivalence agreements between importing and exporting countries; and enhance organic agricultural systems locally and globally.

Governments can use these guidelines to develop their own national organic agriculture programmes, tailored to their specific needs. Most national standards, such as those of Argentina, European Union countries, India, Japan, Tunisia and the United States of America, set forth regulations which are legally binding. In the major organic markets, certifiers are approved and supervised by the national governments.

Organic Agriculture Regulations

The rapid growth of organic agriculture production and trade has been accompanied by an increase in governmental regulations. Regulations establish rules for organic farmers and processors through standards, give credibility to certification bodies through approval and supervision, protect consumers against mislabelling and fraud through conformity and surveillance, and eventually can facilitate trade through equivalence agreements with export countries. Regulations usually include provisions that further promote organic agriculture such as support to research, production and marketing.

At the regional level, the European Union (EU) legislation on organic production, Regulation 2092/91, was adopted in 1991 and is amended yearly. The Regulation sets mandatory standards for organic agriculture with rules on production, labelling and inspection. The aim is to ensure fair competition among organic producers and to gain credibility among consumers. EU Member Nations support the Regulation with domestic legislation, creating their own inspection and certification schemes. Member Nations are required to meet the Regulation's minimum requirements, but they may enact stricter legislation.

In the last decade, many developing countries and countries in transition used Regulation 2092/91 as a model for designing their own national legislation, to assure they will meet EU import requirements and thus enhance their export opportunities.

Along with the European Union, Japan and the United States of America are the main organic markets. Japan began implementing its organic agriculture regulations in 2001, the United States of America in 2002. While similar in many respects, there are also several dissimilarities, both in coverage and substance. This means that exporting countries must be aware of the variations in specific requirements in order to access the different markets. Of course, this need to comply with several regulatory systems increases costs at the production, certification and accreditation levels.

According to a survey conducted by IFOAM in 2002, about 40 nations have completed the process of enacting their organic legislation, including detailed standards and institutional arrangements. Another 15 or so nations are in the process of developing organic legislation.

Government regulations have grown steadily, but often in an uncoordinated manner.

The resulting variation in production, inspection and certification systems hinders farmers' access to regulated organic markets and the proliferation of organic labels can be confusing to consumers. Compliance with foreign organic standards is not always suitable to agroecological conditions of exporting countries. Inspection and certification costs may be out of reach for small producers.

Alternative systems, more relevant to small farms and local marketing, are being developed by the private sector. Many countries have developed Internal Control Systems and in the United States of America, a Certified Naturally Grown labelling programme has been created. Standards, inspection, certification and accreditation (or the organic guarantee system) ensure credibility. There is need for a more harmonised approach to regulate the sector, to ensure that organic rules and standards are designed and applied consistently and fairly, taking into account all stakeholders' interests. Flexible requirements to establish equivalency among regulatory systems is the main challenge ahead for the organic agriculture community.

Organic Production of Tropical and Sub-Tropical Fruits

The modern organic agriculture movement evolved in developed countries, mostly in temperate regions. Now, with growing interest in organic cultivation as a management method for agricultural production in tropical and subtropical countries, greater attention needs to be given to developing standards and guidelines for organic agriculture applicable to tropical products and ecosystems.

Motivation for converting to organic management can include market demand, lack of access to inputs, environmental concerns, social concerns or personal philosophy. In addition, farmers' production targets need to be considered in terms of both quantity and quality, weighing the costs of production against the expected returns. Clear and objective information about both the potential and the constraints of organic agriculture is essential for any farmer contemplating conversion.

Export marketing opportunities for developing countries include those organic horticultural products that are not produced domestically in temperate countries, such as spices and tropical fruits and vegetables. Tropical and subtropical fruit production takes place in settings ranging from traditional home gardens to large plantations, making it difficult to prescribe organic husbandry practices that are suitable for all. Nonetheless, any fruit growers contemplating organic conversion need to have a clear understanding of aspects of organic production and marketing before taking investment decisions.

A challenge for organic fruit producers is the maintenance of economic feasibility while complying with organic standards. This means dealing with crop-specific problems without using substances prohibited by organic standards. Alternative approaches are needed, for instance, to control ripening of pineapples without chemicals, or to control banana or citrus pests and diseases.

Furthermore, conversion to organic agriculture requires innovative solutions that guarantee purity of seeds and planting material and provide assurance against the risk of pest and disease spread. Some of these problems might have partial local solutions while others require specific research to identify or generate the appropriate permitted management practices.

Organic horticulture makes optimum use of crop residues, green manure and biological nitrogen fixation, but this does not replace all nutrients removed by crops.

Nitrogen supply needs to be supplemented by the provision of other essential plant nutrients that usually cannot be replenished sufficiently by organic matter in acid tropical soils.

Considering the nutrient demand of perennial fruit crops over time, it is important to have adequate methods to maintain soil fertility, especially levels of phosphorus, potassium and micronutrients, in such a way that yields can be sustained. There is a strong need to set up networks for sharing empirical experience and scientific results of various organic methods of fertilisation, in order to get a more complete picture of long-term successes that can be shared with organic farmers.

Pest Management

Pest control in organic agriculture begins by making sensible choices, such as growing crops that are naturally resistant to diseases and pests, or choosing sowing times that prevent pest and disease outbreaks. Careful management in both time and space of planting not only prevents pests but also increases population of natural predators that can contribute to control of insects, diseases and weeds. Organic agriculture requires informed decision-making about a range of pest management practices which are being developed. This decision-making and field management process, known as Organic Pest Management (OPM), is an outgrowth of Integrated Pest Management (IPM).

Although the latter allows usage of synthetic inputs, both systems emphasise prevention through such methods as: growing resistant varieties of crops; growing in the proper season for the variety; improving soil health to resist soil pathogens and promote plant growth; rotating crops; encouraging natural biological agents for control of disease, insects and weeds; using physical barriers for protection from insects, birds and animals; modifying habitat to encourage pollinators and natural enemies of pests; and using semi-chemicals such as pheromone attractants to monitor and trap pests.

For curative practices, OPM limits itself to those products approved by certification bodies. But even among these products, there are differences in costs, applicability and access to be considered by farmers. There has been a great deal of research into OPM, and farmers' networks and organic agriculture associations encourage exchanges of experiences. Products are approved based on their low risk for environmental and health impact as well as effectiveness in controlling target pests.

It is important to point out that some products allowed for OPM may be hazardous to applicators. For example, red chili mixtures or garlic oil which are used as pest repellants can be irritating to the skin. Applicators should have proper protective gear and application equipment. At this point, most organic products, such as larvae of pest predators, are imported by developing countries.

There is a need to promote local production to reduce import dependency. Wider accessibility and low tariffs should be encouraged. Farmer education in OPM can be improved with better written materials, better field courses and, of course, exchange visits to meet with and learn from successful farmers.

IPM programmes use "farmer field school" methods to teach ecological approaches to pest management. Similar types of group study are applicable to OPM, especially when combined with testing new methodologies, crops and marketing. Study groups and mutual support groups could also be used for peer review of practices within a specific organic labelling scheme.

Horticulture and Post-Harvestc Operations

An organic label on a product represents more than proof of organic production. It shows that the entire food chain—from the farm to the table—adheres to organic standards. The steady increase in market demand for organic fruits and vegetables in recent years has presented an export opportunity for farmers in developing countries who practice low-input agriculture. However, they must overcome significant hurdles in order to benefit from this opportunity.

Apart from regulations that apply to the production of organic produce, certain post-harvest activities need to be modified to comply with organic regulations. Many post-production operations for organic produce are identical to non-organic production. Although some root, tuber and bulb crops require a curing period at ambient or elevated temperature to promote wound healing and ensure optimum storage life, there are no specific requirements for curing, storing or transporting organic produce. Most markets require strict attention to the size, grade, quality and maturity of produce, whether it is organic or not.

Fruit and vegetables must be cleaned and graded to comply with these regulations. Although all types of packaging are authorised, there is an expectation that careful thought will go into choosing packaging with regard to its environmental impact. Degradable packaging material is increasingly requested by conscientious consumers.

All harvested fruits and vegetables should be placed as soon as possible in a storage area that is kept at the appropriate temperature. However, organic products need to be stored and transported with proper identification and physically separated from non-organic products. As for processing, freezing is the only processing method that keeps produce in a state similar to the fresh crop.

Organic foods can also be processed by drying; with the use of approved processing aids such as: ascorbic acid, citric acid, tartaric acid and salt; blanching with high temperatures to destroy microorganisms; pasteurising to destroy microorganisms that could contaminate the product after blanching; and with heat treatments that conserve products by destroying or inactivating enzymes and killing microorganisms.

Juice is an ideal product for the organic market. It offers a simple and natural way to process fresh produce, preserves the majority of vitamins and minerals, and largely resolves the problem of storage. Canned produce must be prepared in a way that retains, as closely as possible, the characteristics of fresh produce. Other forms of processing include conservation with sugar which is principally used for fruit jams and purees, and by fermentation which is a chemical change caused by enzymes, bacteria or microorganisms. There are very few approved organic post-harvest treatments for pests and diseases. Hot-water (45-55 °C) immersion, steam and forced hot-air treatments are sometimes used as organic control methods after harvest.

Grasslands

Organic production methods and certification standards applied to grasslands contribute to the maintenance of the ecosystem, add value to grasslands products, assure income generation to their users, and promote animal welfare. Yet organically certified and managed grasslands account for only a small fraction of the grasslands worldwide.

Grasslands, which are immensely different in their climatic needs, physical conditions, cultural and historical roots and economic situations, contain 24 percent of the world's vegetation and cover about 4.6 billion ha of which 9 million ha are organically certified. Grasslands are of prime importance in global carbon budget, maintenance of flora and fauna biodiversity, control of desertification, maintenance of soil fertility, and the living conditions of the population.

Grasslands can be natural, made up of indigenous species and managed with few external inputs. Or, they can be temporary, used in rotation, alternating between cropland and grassland and intensively managed. Natural grasslands produce with very few or no external inputs, maintain soil fertility and ground cover, and contribute to the maintenance of biodiversity and ecosystem functions. Most natural grasslands are relegated to the more marginal lands of the South American pampas, the Eastern European and North African steppes, high mountain areas, the Sahelian, Soudano and savannah zones of Africa, and North American prairies.

In terms of meeting organic objectives, temporary grasslands offer a number of advantages including easier maintenance of high clover content, easy exploitation of residual fertility by arable crops, minimized weed development and better possibilities to provide regular, clean worm-free grasing. However, establishment costs are high.

Organic farms in regions with climate, soil type and topography well suited to cropping are likely to have a mixed crop/grassland system, while permanent grasslands are more frequent in fragile environments and low potential areas. Because organic standards restrict crop and livestock inputs, farmers need to employ preventative approaches to crop and livestock husbandry such as crop nutrient supply, weed control and animal health if they are to build and maintain a viable organic system. Well-managed grasslands used for extensive grasing, such as those in Australia and Argentina, offer opportunities to add value to livestock products on the market through organic certification.

Implications of the organic standards on grassland management

Organic Restrictions	Implication for Grassland Management
No soluble nitrogen	Emphasis on legumes
Minimal use of mineral phosphorus and potassium farm	Emphasis on efficient nutrient cycling within the
No synthetic herbicides	Design management to minimize weed invasion
Minimum amount of purc-feed, minimum proportion of forage in ruminant diets, prohibition of certain high protein feeds	Maximize livestock nutrient supply from forage produced on the farm, especially high protein forage
No routine preventative use of veterinary medicines	Manage grazing to minimize build-up of parasite and disease problems

The Grassland and Pasture Crops Group of the FAO Plant Production and Protection Division is looking into alternatives for certification in communal pasture systems and is preparing "Guidelines on organic pasture production in arid and semi-arid communal rangelands".

2

Organic Agriculture

The Codex Alimentarius Commission defines organic agriculture as "a holistic production management system which promotes and enhances agro-ecosystem health, including biodiversity, biological cycles, and soil biological activity. It emphasises the use of management practices in preference to the use of off-farm inputs, taking into account that regional conditions require locally adapted systems. This is accomplished by using, where possible, agronomic, biological, and mechanical methods, as opposed to using synthetic materials, to fulfil any specific function within the system".

All definitions of organic agriculture converge in recognising that this food system is based on ecological principles and that it relies on using adaptive management for building long-term farm productivity. As such, the principles behind organic agriculture are in line with the Ecosystem Approach advocated by the CBD (Decision V/ 6) and its application focuses on functional relationships and processes within ecosystems; enhances benefit sharing; uses adaptive management practices; carries management actions at all level, and where a legislative framework exists, it ensures intersectoral cooperation.

For practical purposes, organic agriculture guidelines and norms specify minimal requirements to label foods as "organic". These requirements contain mandatory and voluntary practices for organic management. Therefore, throughout the world, the contribution of organic agriculture to biodiversity has different levels of achievements, depending on whether the mandatory requirements only are implemented to obtain certification or whether the broader voluntary requirements are also adopted.

As such, biodiversity achievements of organic farms range from pollution reduction (of systems based on agro-chemical substitution) to fully-fledged holistic systems that harbour semi-natural habitats and a wide range of biodiversity.

Agricultural Pollutants

Building of soil fertility and avoiding the use of synthetic agricultural inputs (including fertilisers, pesticides and genetically-modified organisms) is common to all organic agriculture norms. In order to ensure the agro-ecosystem productivity, a set of practices are adopted. The main strategies and their benefits on biodiversity are outlined below.

Biological control and least-toxic approaches to pest management. Pest management in organic systems is achieved through a number of general and site-specific physical, biological and accepted biological pest control methods. The primary strategy for pest management in organic agriculture is preventative, where the use of sound soil fertility management serves to improve the physical, chemical and biological properties of soils, leading to elevated soil fertility and optimal conditions for plant growth.

Crops grown in healthy soils tend to be more resistant and resilient to pest and pathogens and require little to no applications of pest control materials. The reduction or absence of synthetic soil fumigants, insecticides, herbicides, fungicides, etc. serves to prevent exposure of agricultural and surrounding ecosystems to the toxic effects of pesticide pollution. Because pesticides have a direct suppressive effect on many organisms within the agro-ecosystem, the elimination of pesticide use removes a major obstacle to the diversification process of agro-ecosystems. In protected area landscapes, the avoidance of pesticide use by organic farmers is conducive to wildlife conservation.

The use of natural soil amendments: The rational use of soil amendments is a critical aspect of organic system management. The use of soil analysis and nutrient budgets matches agricultural inputs with the nutrient demands of crops in order to make more efficient use of available resources and avoid the pollution problems resulting from the over application of agricultural nutrients. The use of naturally occurring mineral soil amendments serves to supply essential plant nutrients while reducing nutrient leaching and/or runoff.

Compost is used to improve and maintain soil organic matter levels and, when combined with an appropriate legume/grass cover crop and the incorporation of crop residues, it helps support larger and more diverse population of soil organisms. Higher soil biodiversity has shown to increase the rate of nutrient cycling, improve soil aggregation and aggregate stability and improve the disease suppression of agricultural soils.

Use of adapted genetic resources: While there are no, as yet, specific requirements for the use of genetic resources in organic certification systems, the restriction on use of synthetic inputs and genetically-modified organisms leads organic farmers to use and regenerate locally adapted landraces and genotypes resistant to climatic stress and diseases. It is also economically advantageous for organic farmers to use traditional varieties that are productive under low-input conditions, that can survive adversities and that can grow in marginal environments. The organic market often valorises endangered under-utilised and threatened species, including varieties and breeds that have specific nutritional characteristics (e.g. gluten-free crops). Research on selection, improvement and distribution of open-pollinated varieties is at early stage of development.

Polycultures (intercropping and strip-cropping): Polyculture is the intentional cropping of multiple species within a farm, in a planned spatial sequence. Intercropping is a direct way of increasing the diversity of an agro-ecosystem by introducing much more structural and chemical variation to smaller areas of the agricultural landscape and thereby encourages more complex interactions between crops and other organisms. Strip cropping, where strips of one crop are planted next to strips of another is a less management intensive form of multi-cropping.

When compared to monocultures, polycultures have consistently shown lower population of pest insects and weed population. Increases in the variety of food sources, the creation of micro-habitats as well as the increased difficulty of pest population within polycultures to locate dispersed patches of crop hosts serve to limit the growth of pest population and stabilise predator-prey and parasitoids-host population dynamics.

Crop rotation: Crop rotation is the intentional planting of crops in different areas within a farm, in a planned temporal sequence. Crop rotations serve to provide new above- and below-ground habitats, as each new crop has a distinct chemical and biological make-up, introducing new vegetation types to the landscape and eventually, crop residues to the soil ecosystem. Different crop residues promote or inhibit different soil organisms which may have inhibitory or growth promoting effects to subsequent crops or pests.

By interrupting the continuous presence of a crop host, crop rotation serves to break the build-up phase in the cycles of weeds, insects and diseases, thus eliminating the need for pesticide application. Fallow periods, where ground is left uncultivated for an extended period of time (a few months to one year), allow a limited amount of secondary succession to advance and hence, the recovery of the diversity of both terrestrial and below-ground species.

Cover cropping: Cover cropping is a type of non-crop cover grown specifically for soil improvement purposes. Both annual and perennial cover crops may be used, with varying benefits to above- and below-ground biodiversity. Cover crops may provide a physical temporary habitat for many different species of ground-nesting birds and small mammals as well as nectar and pollen sources for many species of insects.

The habitat value of cover crops varies by species and variety; therefore, cover crops must be carefully selected to meet specific management objectives. Cover crops root system improves water penetration and prevents soil erosion. Cereal cover crops planted as an over-wintering rotation may also capture plant available nitrogen, thus preventing nutrient leaching into sensitive water ways.

Cereal and legume cover crop mixes can be an important source of organic matter when incorporated into the soil. The use of perennial cover crops in orchards and vineyards is an effective means of enhancing the biodiversity and productive capacity of cropping systems and avoiding the labour and environmental risks associated with herbicide use. In Mediterranean climates, economic and agronomic advantages may be gained through the use of native perennial grass species in vineyards having a summer dormant growth cycle. As a result, they provide the advantages of a perennial cover crop without the disadvantage of excessive competition or water use.

Minimum tillage: Although reduced tillage is not a required practice in certified organic agriculture, it is a practice that is consistent with the objectives of organic system management and encouraged by most certifying programmes. The minimisation of tillage will often lead to increased earthworm abundance and activity, increased population and diversity of decomposer organisms and an associated increase in the organic matter content and aggregate stability of soils.

When compared to conventionally tilled soil, low and no-till systems exhibit improved nutrient and water holding capacities, improved nutrient cycling and more desirable physical properties. The enhanced biodiversity of these systems lend themselves to improved cropping performance and greater resistance and resilience to various kinds of disturbances. Farming systems using no and low-till approaches are also less likely to incur soil erosion and require less inputs of energy.

Creation of Habitats

Although biodiversity conservation is implicit in most national and international organic standards, explicit biodiversity conservation requirements for non-agricultural purposes remain underdeveloped at this time. In recognition of the need to establish standards for biodiversity conservation in organic agriculture, the 2002 revision of the IFOAM Basic Standards for Organic Production and Processing outlines a set of recommended practices that may serve as a foundation for establishing certification standards for biodiversity conservation in agriculture.

These standards' section on "Organic Ecosystems" requires measures for maintaining and improving landscape and enhancing biodiversity quality as well as prohibition on clearing of primary ecosystems. General recommendations are made on the maintenance of an appropriate portion of the farm as wildlife refuge habitat through, inter alia: the establishment of hedges, hedgerows and ecologically diversified field margins as well as areas with ruderal flora and wildlife corridors that provide linkages and connectivity to native habitats.

Agro-ecosystem heterogeneity (and therefore biodiversity of the landscape) may be increased through the conservation, management and/or restoration of various on-farm landscape features such as soil, water and native non-crop vegetation. The techniques outlined below require varying degrees of modifications of current agricultural practices. Not all organic systems implement habitat enhancement practices and there is scope for improving organic standards for biodiversity conservation.

Minor farm habitat enhancement strategies: A number of on-farm management techniques may be used to protect or enhance on-farm habitat for wild biodiversity. Leaving natural snags and the erecting of artificial perches or houses for native song birds, raptors and bats may provide on-farm roosting, nesting sites and refuges to support population of wildlife. Where mechanical equipment is used for on-farm vegetation management, harvest or clearing of land, the postponing of such operations until after ground-nesting fledge lings have emerged may significantly decrease nesting losses.

Development of wetlands and multipurpose farm ponds: The extensive loss, worldwide, of natural wetlands makes agricultural wetlands most important to wildlife. In particular, flooded rice fields have an enormous significance in creating temporary wetlands for reproduction and feeding of migrant birds and other wildlife.

These important agriculture wetland functions are best addressed by organic management that avoids using synthetic pesticides and fertilizers. In other environments, the development of farm ponds may serve to attract diverse wildlife and may be an economical source of water for small farmers.

Planted with native aquatic vegetation, farm ponds provide food and cover, resting and breeding areas for wildlife. Migratory waterfowl may be attracted to these ponds for food resources and may become occupied year around with salamanders, turtles, shorebirds and certain species of ducks. Wide ranging animals may also use small farm ponds for watering holes. In agricultural areas reliant on furrow irrigation, tail water pond systems are an effective way of recycling irrigation water, trap and filter sediments and provide year around water habitat for wildlife.

Leaving strips of unharvested crops for wildlife: Leaving some unharvested portion of fields for wildlife forage has proven a highly effective means of maintaining or enhancing population levels of many species. Food plots provide over winter food for wildlife which is particularly useful during times of stress or in areas when habitat modification has been significant. Location, size and spacing are several factors to consider when planning food plots. Although plot size and shape may vary according to the type of planting and the intended wildlife use, it is commonly recommended that wildlife lots be scattered over the entire property if possible. Corn is the most common forage plant for wildlife, but annual rye, millet and buckwheat are also beneficial for both mammals and waterfowl. In temperate climates, perennial crops such as clover, alfalfa and other legumes can be planted to provide food for turkeys, songbirds, rabbits and deer in the summer.

Native hedgerows: Intentional plantings of complex assemblages of native trees and herbaceous and woody perennials along field margins and areas of marginal or highly erosive soils is an effective means of enhancing biological diversity and the production capacity of agro-ecosystems.

Hedgerows introduce greater micro-habitat diversity to agriculture lands and can reduce the velocity of winds and runoff when planted along field margins. As a result, hedgerows reduce soil degradation, increase crop yields by conserving soil moisture, increase livestock productivity by buffering extreme temperature variations.

The functional role of hedgerows is being researched to enhance biological control of agricultural pests through specific hedgerow species assemblages. Complex hedgerows provide shelter and food which serve to elevate the carrying capacity for wildlife in an otherwise simplified agricultural landscape.

They also serve as a corridor for the movement of animals between intact patches of habitat that have been interrupted by agricultural development. When hedgerows are planted between natural ecosystems and agricultural lands, they can also help to reduce the potential impacts of agriculture on native plant and animal communities by modifying winds, moisture levels, temperature levels and solar radiation.

Vegetative buffer strips. Vegetative buffer strips are mixed native vegetation of various widths that are planted along field margins, irrigation and drainage ditches, tail water ponds or between agricultural areas and sensitive native habitats to serve both an environmental protection and biodiversity enhancement function. Being climate adapted, regionally specific native grasses and woody perennials will need only minimal care until they become established (many perennial native grasses and shrubs will also re-seed themselves).

Grass species often have extensive root systems that may serve to prevent erosion and catch sediment and nutrients, thereby filtering runoff and improving water quality. Such forms of vegetation management serve to increase the complexity of vegetation in agricultural landscapes, providing diverse pollen and nectar and micro-habitat for native species of insects, birds and small mammals.

In organic agriculture, vegetative buffer strips also function to protect from the negative impacts of neighbouring conventional farms, as they catch sediment and excess nutrients or agricultural chemicals. Vegetative buffer strips can also serve to reduce the spread of fire from the agricultural areas to the native vegetation. Planting complexes of regionally specific native perennial grasses and other herbaceous and wood perennial species in buffer strips can be an effective and inexpensive way to address the common management problem of weed control in these areas.

Riparian corridors: Due to the fragmentation of habitat common in agricultural areas, the maintenance or re-establishment of corridors between intact habitat remnants is an important component of enhancing the habitat value of agricultural areas. Riparian vegetation connects terrestrial and aquatic ecosystems creating extensive ecotone areas and serve as corridors for the movement of plants and animals between areas of undisturbed habitats.

Riparian corridors are considered to be one of the most diverse and complex terrestrial habitats on earth. Riparian corridors play a critical role in filtering sediments, providing organic matter to aquatic organisms, regulating water temperatures and dissolved oxygen levels, sequestering excess nutrients and absorbing surface runoff and preventing or reducing flood damage.

Loss of riparian vegetation can significantly alter the function of these systems leading to loss of migration corridors, colonisation of invasive exotic plant and animal species, stream bank erosion, alteration of streambed habitat, reduced ability to assimilate chemical and soil losses from agriculture lands and associated degradation of water quality. Maintaining the quality of the water mitigates the need for watercourse clean-up, and encourages wildlife and aquatic life.

The UK Soil Association standards for riparian buffer zones for water quality protection recommend a minimum of 2 metres buffer of riparian vegetation on each side of a stream. In Northern California agricultural areas, a minimum of 30 metres on both sides of the stream is often recommended to provide significant ecological values such as wildlife corridors and biological control of agricultural pests.

Buffer zones: Buffer zones link protected areas to one another through the preservation of habitat. By definition, these areas are intended to remain under wild cover or be managed in such a way as to ensure that human land use is compatible with the maintenance of a high degree of biological connectivity. Buffer zones are used in protected areas to delineate areas where human activities such as grasing, non-extractive agroforestry, under-story cropping, hunting and gathering may occur.

Depending on the species selected and how the area is managed, buffer zones may serve to reduce the pressures placed on intact native vegetation for the production of foods and fuels for human consumption. Buffer zones play an essential ecosystem function by allowing for the dispersal of plants and the movement and migration of animals between intact habitat types. To be effective, buffer zones must offer suitable habitat to wildlife and therefore the preservation of such environments free of pollutants along such pathways is of critical importance.

Under-story cropping: Under-story cropping is one the most effective ways to conserve native habitats and biological diversity of tropical forest areas that are used for agricultural production. Many traditional cropping systems integrate crops into the under-story vegetation of thinned primary or old secondary forests for the production of coffee, cacao, yerba mate and some tea varieties. Those traditional systems, which minimally modified the native habitat, are currently replaced by intentionally planted polyculture systems composed of native and non-native species that serve various ecologic, economic and cultural functions.

In the last decade, the biodiversity value of shade cropping and its role in the sustainable management of protected area landscapes has emerged and a return to traditional polycultures is being achieved through organic agriculture and other biodiversity-friendly approaches. Under-story cultivations present enormous opportunities for the conservation and enhancement of protected areas and buffer zones while securing food and livelihoods for millions of people living in tropical forests.

Integral Strategies

Landscape Planning

In addition to on-farm management through organic and habitat enhancement practices, the success of biodiversity conservation in protected areas will depend upon how well individual farms are integrated into the wider landscape. Through proper planning at a landscape level, natural and managed areas may be suitably integrated into the landscape in order to reconcile human activities with the goals of biodiversity conservation.In order to best integrate the multiplicity of objectives within protected areas, land managers need effective planning tools to properly manage multi-use areas for the benefit of all interests.

Landscape ecology is the study of environmental factors and interactions at a scale that encompasses more than one ecosystem at a time. Because landscape ecology helps to understand how different parts of the landscape mosaic are formed and how they interact, it may provide a basis for the effective management of agricultural landscapes within protected areas. Landscape planning provides a tool for the integration of agro-ecosystems with the broader landscape in a manner that minimizes habitat fragmentation and that confer the beneficial aspects of biodiversity to agro-ecosystems.

The tools commonly used in landscape planning are mapping, and where available, aerial photography and geographical information system analysis. Such tools provide a means through which historical, current and proposed landscape features and land-use activities may be contrasted to specific conservation goals. With this information, recommendations for changes in land-use practices and agro-ecosystem design may be made that serve to best integrate agricultural ecosystems and other human land uses into the broader landscape.

In doing so, impact to critical habitat and sensitive areas may be mitigating or minimized and greater coherence between on-farm habitat and the surrounding landscape may be made. Such management tools may also serve in monitoring the effects of specific land-use practices in order to gauge the effectiveness of interventions.

Landscape ecology and planning may be used at various scales. At a larger scale of analysis, landscape planning may be used to determine the most appropriated placement of agriculture and other land-use activities when attempting to reconcile human livelihoods and habitat needs for biodiversity conservation.

By looking at the landscape of which individual agriculture ecosystems are a part, key ecological features may be identified for use to determining appropriate on-farm habitat restoration efforts that may best serve to re-establish continuity of habitat types and thereby reduce habitat fragmentation. If such ecological features are managed successfully between a number of contiguous farms and wild surroundings, a landscape that is more compatible with the needs of a more diverse number of plants and animals may result.

Agro-ecotourism

Several CBD references mention the link between ecotourism and agriculture: "ecotourism provides for full and effective participation and viable income-generating opportunities for indigenous and local communities". "In order to contribute to the sustainable use of diversity through tourism, there is a need to implement flexible mix of instruments such as integrated planning... ecolabelling...". "In some areas, low-input and small-scale agricultural activities that result in both an attractive environment and the maintenance of high levels of biological diversity can offer an opportunity for tourism. Sale of products derived from sustainably harvested natural resources may also provide significant opportunities for income-generation and employment".

Ecotourism: Under the right circumstances, ecotourism has proven to be one of the most effective means to finance biodiversity conservation. In most rich biodiversity areas, actual revenue flows for ecotourism are better than non-timber forest products and biopharmacy, and comparable only to agroforestry.

Because the dominating land use in protected areas and buffer zones is agriculture and forestry, ecotourism is an opportunity for the creation of additional income to farmers/foresters and to generate financial means for the management of protected areas, especially where governmental park management agencies have little resources.

Agrotourism: The symbiotic relationship between tourism and agriculture that can be found in agrotourism is a key element of an environmentally and socially responsible tourism in rural areas. Rural hospitality offers new employment and income generating opportunities for rural population, including agrotourism as expression and cultural exchange of agricultural practices, artistic heritage and craftsmanship and culinary traditions. Agrotourism may take several forms: holiday farms, farmhouse bed-and-breakfast, farm camping, mountain resorts, equestrian centres and other forms of rural accommodations. Such facilities are an innovative payment system for environmental services generated on and around agricultural lands.

Agro-ecotourism: While ecotourism is nature-based and agrotourism is farm-based, agro-ecotourism is a combination of both. The rural landscape, usually a combination of wild and agro-ecosystems, is the most important resource for tourism development. It is obvious that a diversified agricultural landscape, with semi-natural habitats, has a greater aesthetic and recreational potential over uniform, degraded and/or polluted agricultural areas.

In Europe, agri-environmental policies often promoted organic agricultural activities as a most effective means for landscape conservation: for example, the European Union Life Environment project run by the French Federation of Parks and Reserves adopted extensive animal husbandry to prevent the negative impacts of unmanaged forests on some botanical meadow species and to promote a landscape quality attractive to tourists.

Examples from the Alpine Region showed that agriculture maintained an ecological value much more attractive to tourists than areas where agriculture activities were extremely reduced. Tropical countries that harbour extraordinary biodiversity have an untapped potential for generating tourism business around biodiversity-rich farms.

Eco-organic tourism: When agro-ecotourism evolves around an organic farm, it is referred to as eco-organic tourism. The valorisation of specific elements of the agro-ecosystem landscape offers an additional economic resource for environmental protection. Conversion to organic management in agricultural areas and the development of connected activities such as tourism are increasing.

When farms are organically-managed, they increase the motivation for tourists' visits. New tourist expectations have enhanced the quality of the supply such as diversified farm landscape, environmentally-sound farmhouse architecture and local/typical gastronomy.

In many industrial countries, protected area landscapes including farmland experience land abandonment. In this context, small-scale agriculture can become economically viable if quality products could be marketed and income is supplemented by tourism activities, especially in areas rich of biodiversity and history.

While the establishment, by local authorities, of land protection systems was historically opposed by farmers unions, there is a recent growing awareness that a reserve/park increases "green" tourism opportunities and that visitors are increasingly sensitive to quality, both in ecological and gastronomic terms. This trend favours organic farmers because they can easily meet these new tourists' demands.

Different studies in the European Union demonstrated the ability of organic agriculture to create attractive landscapes. Eco-organic farm practices and activities which benefit the environment while rewarding farmers include: accommodation in buildings renovated/built according to ecological architecture; on-farm consumption or selling of organic foods and beverages; educational programmes and training, and sensitising guests on rational use of natural resources. A 1999 survey in Italy reported that eco-organic holiday farms performed the following activities: 35 percent arrange visits to nearby protected areas; 30 percent plan naturalistic didactical activities; 24 percent set up didactical and demonstrative laboratories on organic agriculture and the environment; and 11 percent offer visitors instruments and equipment for the observation of fauna and flora.

Ecolabels are important marketing instruments for agro-ecotourism in general, because price premiums encourage farmers' commitment to the conservation and maintenance of biodiversity. The most well-known forms of certification include organic farming operations, organic and specialty foods and forest stewardship products.

Organic certification of farmhouse structure and facilities is less known but where implemented, it attracts more environmentally-conscious tourists. Requirements for such farms include: organic agriculture production; naturalistic and didactic activities; natural resources tutorship ; recreational green areas; ecological buildings ; tourist offers ; gastronomic offers ; and sustainable transportation facilities.

Protected Area Management

The importance of ecoforestry and Sustainable Forest Management (SFM) to protected area management is that these approaches to forestry provide opportunities to generate financial and livelihood benefits to local communities which give them an incentive to support conservation initiatives such as protected areas. Ecoforestry has its roots in environmentalism, whereas SFM originates in production forestry.

Ecoforestry

Self-sufficient and stable human communities can grow strong from a sense of place, recognition of interdependence and respect for the forest. Some key values of ecoforestry are ecological integrity, community vitality and economic opportunity. Its methods are specific to bioregions and forest ecosystems. Ecoforestry favours value-added manufacturing and local jobs by providing a continuing, diverse and local supply of forest products.

Ecoforestry is a low-impact approach to forest management that maintains a fully functioning forest within the natural historic range of spatial and temporal variability. Its practices favour native tree and plant species which provide for the needs of wildlife and their habitats.

Examples of the basic ecoforestry principles and practices are:

— observing the structure, function, composition and natural changes of forest ecosystems, learning from these and using management practices that mimic them;

— preserving the natural diversity of ages, heights and species of trees;

— protecting wildlife and their habitats;

— managing for logs as just one possible product among many non-timber forest products;

— focusing on what should be left after harvesting (in order to keep the ecosystem functioning) rather than focusing on what one takes;

— using low-impact logging systems;

— the volume of trees removed is less than the forests growth rate in order to provide for forest structures such as standing and downed dead trees;

— promoting natural regeneration;

— appreciating all the other forest values (aesthetic, spiritual, genetic, recreational, protective...) at least as much as the monetary value of marketable products;

— the precautionary principle: when in doubt, as to whether a potential action in the forest may be harmful to the ecosystem or not, don't do it!

Certification is being used as a mechanism for promoting ecoforestry, with the Forest Stewardship Council being the preferred approach for the Ecoforestry Institute. During the last ten years, the Forest Stewardship Council has certified over 30 million hectares of forest and other 70 million hectares have been certified under other schemes.

Although certification initially focused on large-scale industrial logging, community-owned forest enterprises and small holders, including indigenous people, are now being certified.

Ecosystem Based Management is another way in which ecoforestry is being characterised and this has been taken up on a large scale. The "Great Bear Rainforest" agreement announced by the provincial government of British Columbia in April 2001 between environmental organisations, First Nations communities, and forestry companies working on the north-central coast heralded an environmental coup, the reverberations of which will challenge the British Columbia forest industry.

The agreement signalled the success of an international boycott campaign aimed at forestry companies singled out for environmentally degrading operations. As well, the agreement protects huge tracts of land from logging, including 96 458 hectares to preserve essential habitat of a rare white subspecies of black bear popularly known as the "Spirit Bear".

But the real environmental victory forged in the agreement lies in the announcement that all development activity in the plan area, which covers about 4.8 million hectares of marine, foreshore, and upland area on the north-central coast, will be based on the principles of Ecosystem Based Management.

Concept of sustainable forest management

The guiding objective of the Forest Principles is to contribute to the management, conservation and sustainable development of all types of forests and to provide for their multiple and complementary functions and uses. Principle 2b specifically states that "Forest resources and forest lands should be sustainably managed to meet the social, economic, ecological, cultural and spiritual needs of present and future generations." It goes on to specify that: "These needs are for forest products and services, such as wood and wood products, water, food, fodder, medicine, fuel, shelter, employment, recreation, habitats for wildlife, landscape diversity, carbon sinks and reservoirs, and for other forest products." And that: "Appropriate measures should be taken to protect forests against harmful effects of pollution, including air-borne pollution, fires, pests and diseases, in order to maintain their full multiple value".

Although the Forest Principles form a "non-legally binding statement of principles", they bear the marks of a negotiated text with some repetitions and passages with ambiguous or very general text and are, in places, focusing on guidance for the establishment of an enabling framework for SFM, rather than specific guidance for application at the field level.

Nine ecoregional forestry initiatives or processes involving 149 countries, whose combined forest area equals 97.5 percent of the total forest area in the world, have been established since 1992 with the aim of translating the concept of sustainable forest management into practice. They continue to meet on a regular basis to further refine the concept through the development of criteria—or elements—defining SFM and sets of indicators for each of these aimed at facilitating monitoring of progress in the practical application of the concept.

Although evolving independently, these ecoregional processes are conceptually similar in objectives and overall approach and have shared information and experiences resulting in a convergence as regards the main elements constituting SFM.

The 2003 International Conference on the Contribution of Criteria and Indicators for Sustainable Forest Management (CICI) was held in Guatemala gathering representatives from all of the above processes, international organisations (including NGOs), government officials and experts in the field. Participants at this Conference agreed that SFM comprises the following seven common thematic areas:

— extent of forest resources

— biological diversity

— forest health and vitality

— productive functions of forest resources

— protective functions of forest resources

— socio economic functions

— legal, policy and institutional framework.

Of these, four are related to the environmental aspects of SFM and the remaining three to the social and economic aspects—the two other "legs" of SFM.

As would be expected, the indicators vary widely among initiatives owing to differences in forest types and environmental, social, economic, political and cultural conditions. National-level criteria and indicators are being complemented by the development and implementation of criteria and indicators defined at the forest management unit level in a number of the processes as well as by other actors such as NGOs and the private sector.

Challenges for the Effective Management

The main challenge in protected areas is to conserve biodiversity while providing the basis for the social and economic development of local residents. In most parts of the world, biological fragmentation compares with social and economic fragmentation. For biological diversity to be conserved and restored through the agronomic and planning practices outlined above, it must be understood that a considerable number of social, economic and or policy factors must be changed to encourage the adoption of such programmes. Ecological knowledge and economic feasibility are key to the further advancement of initiatives designed to increase the biodiversity of agricultural landscapes within protected areas. More importantly, collaborative management between both local communities and sectoral policy makers is a precondition for success.

Biological fragmentation	Social fragmentation	Economic fragmentation
Species and ecosystem Extinction	Food insecurity	Vulnerability of local economies
Micro-climate alterations	Loss of life forms and practices	Decreased production alternatives
Alteration of vital cycles (land, water, air)	Loss of identity	Homogenisation of local economy
Species break-in	Trust in local capacity lost	Relationship with other economies lost
Landscape rupture and vulnerability	Geographic and cultural uprooting	Resources valued from a strictly economic point of view

Dissemination of Ecological Knowledge

Consumers awareness: To date, few agricultural lands in or around protected areas are managed organically and considerable knowledge (and subsequent policy efforts) will be required to convert current agricultural practices to organic practices. Lack of awareness on the benefits of organic agriculture, especially of policy-makers, remains a main constraint to its promotion and adoption. Increased awareness of both environmentalists and agriculturists of the promise of organic agriculture and connected tourism activities in protected areas is key to the scaling-up of numerous but scattered successful experiences. An increased consumers' awareness of organic agriculture and connected biodiversity conservation practices creates a market demand, which in turn creates market-incentives for farmers to adopting biodiversity conserving practices. Studies conducted to date provide a strong basis to move forward to mobilise market and policy forces on behalf of organic agriculture.

Agro-ecological research: The promotion of food production in protected areas requires a commitment to expand research, which results are needed to direct choices made in policy, funding and markets. The lack of basic knowledge of biodiversity patterns in agricultural landscapes inhibits sound conservation practices. Understanding the complexity of trophic and other ecological relationships that are maintained in a structurally diverse farm require substantial research investments. Agro-ecological research, building on local or indigenous knowledge, is key to provide financially realistic management recommendations for farmers and to gain environmentalists acceptance of the agricultural activities in protected areas.

Restoration ecology is in its infancy and much research is required to provide practical answers on vegetation configurations and ecological processes in the food chain. Reclamation of degraded farmlands located near corridors, riparian habitats or protected areas deserves special attention.

For tropical forest conservation, understory cropping associations deserves targeted research to better understand what mix and density of shade trees best enhances biodiversity and crop production. Assessments of the use of shade trees by forest organisms should be overlaid with information on the silvicultural and agronomic properties of those tree species. There is need to evaluate the impact of different management intensities and landscape settings on the long-term viability of birds, trees, epiphytes, invertebrates or mammals, especially those assumed to be dependent on vanishing wetlands and tropical forest habitats.

In particular, the influence of surrounding land use on agro-ecosystem capacity for harbouring sensitive biodiversity (understudied taxa that are sensitive to anthropogenic disturbances) requires better understanding. Such knowledge is important in developing education and training programmes as well as advisory services to producers and protected areas managers.

Education and training: Multi-disciplinary education and training is key to the adoption of ecological production approaches by both agriculture and protected areas managers. Networking and joint ventures between producers and technical experts increase information flow and mutual learning: alternative forms of educational ventures facilitate exchange on production and on-farm biodiversity management and support training of local communities.

Extension work may reinforce small farmers' knowledge on the need of biodiversity enhancing management for sustainable production. Extension workers could also provide materials and seedlings to build community nurseries as well as cooperative assistance for reaching markets for certified products. Improved understanding of the value of integrating conservation planning with development, coupled with an improved foundation for additional incentives for farmers to grow biodiversity-friendly crops, has demonstrated to be successful.

Continued development of organic standards: Voluntary environmental and social certification schemes exist for agriculture products (plants and animals), fish farming, forest products and products gathered in the wild. Ecolabelling varies from Organic, through the Rainforest Alliance "Eco-OK", Bird Friendly, Predator Friendly, Forest Stewardship, Marine Stewardship to Fair Trade and Ethical Trade labels. These voluntary standards have the potential to both raise consumer awareness of sustainable and equitable land-use practices and offers financial incentive to farmers to raise their management standards.

What qualifies organic agriculture as the preferred management tool in protected areas is the prohibition on synthetic agricultural inputs and the organic inspection and certification system that ensures compliance with defined practices. However, requirements for practices which specifically target biodiversity in organic systems remain underdeveloped. Biodiversity standards for low-input agriculture in tropical forests exist and management principles and recommendations for biodiversity-compatible farms have been established by a number of scientists and institutions.

Although the latter standards are not inconsistent with organic standards, moderate use of agro-chemicals is allowed; however, provisions are made for maintaining a percentage of the farm area under forest, establishing riparian buffer zones, fostering diverse native canopies, creating processing plants that divert pulp and liquid wastes from local streams, and improving livelihoods.

Considering that organic products enjoy an established market demand, a combination of organic standards and biodiversity standards specific to tropical forests would be an optimal mix to improve the ecological and economic performance of agriculture in protected area landscapes. The organic agriculture community has started to move in this direction. Such international standards would provide the foundation for developing standards tailored to local environments, traditions and specific production systems.

Capital Incentives

Off-farm income generation: Organic agriculture has developed in response to market-demand and farmers' innovations and commitment to environmental quality. Off-farm income generating activities such as agro-ecotourism, and eco-organic tourism in particular, provide a promising alternative to both small holders livelihoods and sustainable economic activities in and around protected areas.

Secure land tenure: Market pressures and the security of farmers over a given piece of land will largely determine the degree to which farmers are willing to adopt organic agriculture and other biodiversity conservation practices. The 2-3-year transition period required by organic certification schemes and the associated cost of transition represents a considerable investment for farmers.

Farmers need to have secure and long-term access to lands in order to commit to the adoption of biodiversity conserving practices. For farmers without legal land tenure, tenure could be granted to farmers in exchange for adopting a defined set of conservation land use practices.

In protected areas where land is legally owned by farmers, conservation easements may be used to identify land-use practices that must be maintained in exchange for cash payments from the purchaser. The purchaser is often a local, national or international land trust which holds conservation easements and monitors compliance with the legal agreement. A conservation easement is a legally binding plan that is attached to the deed of the property and transferred to anyone who may purchase the land in the future.

Securing capital incentives: Capital incentives are necessary to transition from strictly economically-driven activities to environmentally and socially sound productive activities. The adoption of technologies and approaches necessary for biodiversity conservation in agriculture include business partnerships, private capital investments and public incentive payments. Market-based and other private and public incentives compensate organic farmers for their stewardship efforts. Special interest business groups can provide a sizeable interested market and thereby secure some level of demand for the product.

While market forces provide a sufficient drive for certified organic products, public support and other biodiversity schemes are of key importance for the start-up phase: during the 2-3 years conversion period, farmers cannot capture prices premiums and do incur yield losses. In particular, payment for ecosystem services should be transferred to farmers who grow crops in forests and other natural systems which are degraded and of great biodiversity concern.

Financial incentives should also apply to buffer zone land holders, with a view to encourage farming practices that harbour dispersing native flora and fauna, cushion the impact of invasive species on protected areas, and stabilise agriculture encroachment at the park boundary.

Considering that 60 percent of terrestrial biodiversity is found on a mere 1.4 percent of Earth land area, support to farmers operating in these biodiversity "hot-spots" is a priority. Hotspots as well as all protected areas are islands where anything except collaborative management is not sustainable.

Collaborative management requires intensive local science (agro-ecology and other ecosystem-based approaches) and a conducive political process. Investments in intensive local science and political processes is a precondition for the effective management of protected areas and biodiversity hot-spots.

Collaborative Management

Participation and partnerships: Landscape management plans for protected areas often involve modifications of land-use practices or land tenure. Local cultural and livelihood needs must be considered along with conservation goals. The use of participatory appraisals of problems and solutions, collaborative management, and participatory monitoring/evaluation schemes are effective ways of integrating community member's interest and involvement and thus, generating long-term support to biodiversity conservation projects.

Community-conserved areas are found in every region of the world and take a variety of forms, including indigenous reserves, community-managed ecosystems, managed landscapes, sacred forests and springs, partnership areas, and many privately or NGO protected lands. However, after a decade of implementation, community-based protected area management is still not widely accepted (or practiced) and more rigorous documentation of opportunities and constraints, including an analysis of effective incentive measures, is required for scaling-up (and sustaining) successful experiences.

Integrated policy and planning of protected areas and buffer zones: Organic agriculture alone cannot meet conservation challenges without landscape planning. Also, the ecosystem services on which organic agriculture depends can only be restored and maintained at the ecosystem level. The closed attempt to realign agricultural policy towards more environmentally sustainable systems is found in the agri-environmental programmes of the European Community; within this policy framework, organic agriculture has played a central role in many countries' policies, together with management agreements for biodiversity conservation.

Globally, the fragmentation of natural habitats and the increasing livelihood needs of millions of people living in and around protected areas impose an integrated policy for nature conservation and agricultural development. Integrated policy, planning and management of protected areas represent a major challenge to the agriculture, forestry, environment and tourism sectors. Policies that reconcile the needs of both rural communities and nature conservation has been addressed on a number of occasions but implementation on the ground remains a major undertaking.

Public sector agencies and civil society institutions involved in environmental and agricultural and rural development, together with local community members, should establish cooperation agreements and plans to evaluate, implement and monitor a holistic approach to economic activities in protected area landscapes. Plans related to buffer zones for internationally supported parks should consider the positive environmental and economic effect of organically managed agro-ecosystems.

3

Constraints of Organic Agriculture

Natural resources management is the result of human interactions with ecosystems. Depending on farmers' skills, agriculture can generate either synergistic or antagonistic effects on both the human and natural environments. The economic contribution of agriculture can only be maintained if its base is preserved and their functions are enhanced. There is also need to redirect efforts TO human skills and social resources to create a favourable environment for agriculture.

Sustainable agriculture development needs a paradigm shift from segregation to integration, that is a move from focusing on single, isolated parts of agricultural systems (disciplinary reductionism) to perceiving all parts in their full context (interdisciplinary holism). It explicitly recognises that it is the users of resources, not the stocks of natural resources, that are the focus of management.

Sustainable Agriculture

Sustainability is an issue of A system's performance over time. In order to "measure" the performance of a system, one must define its boundaries and the relationships of their constituent components: there are internal components and external influences as well as inputs and outputs which cross boundary lines.

In agriculture, a system could vary from the soil system, to the cropping system. farming system, watershed system, and national agricultural system. Since the biophysical and socioeconomic conditions are constantly changing, sustainability is more of an evolving concept than a set of fixed rules. Issues of sustainability must be addressed in relation to values, power relationships, time and space: sustainable for whom? for what? and for how long? Developmental processes could be depicted by classifying resources in the following four broad categories:

— *Human resources:* stock of human knowledge and skills, philosophy, ethics, religion, health, nutrition and education, labour;

— *Social resources:* complexes of norms and codes of behaviour that shape human interactions and persist over time by serving collectively valued goals. It includes formal institutions and informal ones, as well as traditions, formal laws and regulations, democracy, good governance, civic actions, equity, social cohesion;

— *Economic resources:* human-made goods generated via economic activities, including income flows derived from the various assets, technologies and infrastructure; and

— *Natural resources:* non-renewable and renewable natural resources and environmental services, including the self-purifying services of renewables.

Objective of Ecological Agriculture

Ecological agriculture should fulfil the following requirements in order to be environmentally sound, non-consumptive, economically viable, socially just, and culturally compatible, that is:

— meet the demands of local geography, particularly climate and topography;

— enhance and build on natural processes, mainly through organic recycling such as nutrient recycling, nitrogen fixation, and pest-predator relationships;

— increase biodiversity and genetic heterogeneity;

— generate only non-accumulative and biodegradable waste and by-products;

— require low-energy consumption, preferably using renewable resources and recycling more of their wastes and products;

— use local materials and energy;

— require available capital investment and affordable maintenance;

— build on locally understood technologies and use proven techniques to reduce the incidence of failure;

— involve local people in its introduction and development and use existing or easily adaptable skills;

— be accessible to, and self-reliant among farmers and rural people, and adaptable to their management capacities.

Organic Agriculture Systems

Production and post-production parameters include:

— land and water management in mixed plant and animal systems that combine land use, tillage, shelter zones, hedges, pastures, agro-forestry, and water use;

— soil fertility management combines eco-physiology, nutrient cycling, crop rotation, inter-cropping, and generation and recycling of organic matter;

— crop production combines crop selection, propagation, and organic pest and weed control;

— animal husbandry combines appropriate feeding strategy, veterinary practices, and stocking rates;

— farm resources involve sourcing renewable energy, labour and mechanisation;

— production handling involves adequate storing, processing and transportation to minimize losses and preserve quality.

Socioeconomic and institutional "infrastructures" involve:

— regulatory framework (voluntary or mandatory) and supporting policies that include aid schemes and instruments to pay producers (especially small operators), develop diverse marketing structures, promote regional development, provide information and legal definitions;

— standards and norms that govern inspection, certification, labelling and accreditation practices;

— research that defines agenda for organic research, provides links with mainstream agricultural research institutes, promote farmers participation and in-situ research;

— community development that targets farmers associations for self-regulatory mechanisms and national networks;

— market structure including outlets, opportunities and comparative advantages, price and demand, organisations, cooperatives and unions (of farmers, processors, retailers, consumers, and sellers/ supermarkets), production-consumption structures, and marketing networks;

— consumer education related to production processes and products quality;

— consumer spending power to support ecological production.

A shift to organic agriculture brings about significant change:

— first, the composition of the inputs changes. Together with a reduction in the use of synthetic fertiliser and pesticides, an increase of other inputs can occur, such as organic material, labour and machinery.

— at the same time, rotations change, affecting yields and yield variability, total production and income (both present and future).

— this, in turns, influences food security, and the natural environment.

— those changes are often influenced by, and influence, social changes within the community.

Farmers would need to evaluate all the above issues to determine their likelihood of success in organic agriculture.

Socio-ecological Analysis

Organic agriculture prospers on diversified farms with complex interactions. It links agro-ecological specificity with socioeconomic diversity. Currently, most policies actively encourage farming that is dependent on external inputs, exogenous technologies and foreign expertise, which consider only the biophysical criteria of an agricultural systems.

Agro-ecosystems are currently defined on the basis of rainfall and temperature, which may be relevant for the evaluation of single crop production opportunities. The complexity of agro-ecosystems is largely based on farmers' decisions, both at individual farmer and farming community levels. Whereas the harvested components are directly linked to the way agro-ecosystems productivity is measured and evaluated by the farmer in the short run, the non-harvested components play a key role in the functioning of agro-ecosystems, its sustainability and long-term productivity, and its function to other parts of society at large.

Organic management includes environmental functions of land use as well as a strong human dimension, namely: farmers' ability to predict future markets, the farming communities' characteristics that determine local markets, market integration, the local economic and social forces that dictate which crops will be grown and when, opportunities for unskilled labour and migration flows, and a host of state variables and attendant dynamic rules and regulations of access.

It is the "human space" where people live and derive resources for a fully reproductive life that delimits "socio-ecological" zones, and therein lies the entry point for sound management of natural ecosystems. Socio-ecological land units are therefore the basis for rural and agricultural activities since they determine land rights and natural resource use. Actions at this level imply an infinite number of decisions, which are taken by users of "local spaces". The challenge is to couple the legitimate needs and interests of individual users with the collective interest for sustainable development, including the health of transboundary natural ecosystems.

Challenges in Organic Agriculture

Ecological Dimensions

The flow of energy (that involves biological and non-biological agents) drives the carbon, oxygen, nitrogen and phosphorus cycles. Nutrients are pumped through the system by the action of photosynthesis and are again made available for recycling by the action of decomposers. Nutrients are constantly being removed or added; adding more natural substances or synthetic materials than the ecosystem is able to handle upsets bio-geochemical cycles.

Environmental services that are vital to agriculture include:

— *Soil forming and conditioning:* A substantial amount of invertebrates (earthworms, millipedes, termites, mites, nematodes, etc.) play a role in the development of upper soil layers through decomposition of plant litter, making organic matter more readily available, and creating structural conditions that allow oxygen, food and water to circulate.

— *Waste disposal:* A succession of micro-organisms occurs in the detritus, involving namely bacteria and fungi as well as detritus-feeding invertebrates, until organic material is finally reduced to elemental nutrients. Ecosystems recycle, detoxify and purify themselves, provided that their carrying capacity is not exceeded by excessive amounts of waste and by the introduction of persistent (synthetic) contaminants.

— *Pest control:* Predation is not just the transfer of energy whereby one organism feeds on another organism but also complex interactions among predator-prey population. If a portion of the prey is not available because of environmental discontinuities (a typical case in agriculture), the self-regulating balance will be dampened.

— *Biodiversity:* An ecosystem stability (or instability) depends on the results of the competition between different species for food and space. Predation ameliorates the intensity of competition for space and increases species diversity. The nature of inter-specific competition and its effects on the species involved is one of the least known and most controversial areas of ecology.

— *Beneficial associations:* Symbiosis of plant roots with mycorrhisal fungi plays a most important role in temperate and tropical forests in absorbing nutrients, transferring energy and reducing pathogen invasions. Parasitism is used in the biological control of insects. Other symbiotic combinations include animal/fish/tree species.

— *Pollination:* 220 000 out of 240 000 species of flowering plants are pollinated by insects.

— *Carbon sequestration:* The capacity of biomass in sequestrating carbon is receiving an increased attention with the aim of reducing (in the long term) climate change. Where no tillage is practised, soil contributes to retaining carbon. As organic agriculture favours minimum tillage (for better retention of water, nutrients, and biodiversity), the carbon retention potential of soils is becoming an important issue.

— *Habitat:* Although by definition, habitats provide shelter and food, many ecosystems have functions often discounted.

Natural resources

Land: Organic farmers build on environmental services by using several techniques to prevent erosion and enhance soil fertility. Soil structure is improved through nutrient mining (by deep rooting crops), improvement of nutrient availability with mycorrhisal and optimal nutrient recycling, specific crop rotation and manuring strategies, and other natural fertilisers. Minimum tillage avoids soil compaction. Integrating trees and shrubs conserves soil and water and provides a defence against unfavourable weather conditions such as winds, droughts, and floods.

Water: Maintaining water quality requires good knowledge of ecological processes and applying farm practices that avoid pollution and use little water. Due to the change in soil structure and organic matter content under organic management, water efficiency is likely to be high on organic farms. Efficient harvesting and use of water also controls salinisation and waterlogging.

Genetic material: Preservation of agro-biodiversity is indirectly addressed by organic agriculture since this form of agriculture relies mostly on endemic biodiversity that is resilient to local ecological stress (e.g. drought). The emphasis of seeds and breeds used in organic agriculture is on local suitability with respect to disease resistance and adaptability to local climate. The availability of suitable genetic material (e.g. GMO-free), its selection, rearing and distribution is often a constraint.

Agronomic Dimensions

Fertilisers and pesticides: Decreasing the use of synthetic fertilisers and pesticides goes together with increasing other inputs. These inputs can be bought or produced on the farm (such as manure), others come in the form of knowledge about actions to be taken. The change in the combination of inputs may change the effectiveness of certain processes, which influence farm output, such as the cycles of water, nutrients, energy and knowledge (inter-generational). Crop protection relies on natural pest controls (e.g. insect pheromones, plants with pest control properties) or by enhancing self-regulation.

Crop rotation: Crop rotation is required under organic certification programmes and is considered to be the cornerstone of organic management. Crop rotation is a valuable tool for weed control, maintenance of soil structure and organic matter, recycling of plant nutrients, contribution to overall species and habitat diversity, preventing erosion (application of cover crops), green manuring, and pest and disease control. Inter-cropping, after-cropping, alley-cropping and mixed cropping are spatial and temporal alternations chosen according to crop and soil-specific cycles and needs. Agricultural pests (fungi, insects, and weeds) are often specific to the host, and will multiply as long as the crop is there.

Manipulation of crops between years (management by rotations) or within fields (strip-cropping) is therefore an important tool in the quest for management of pest problems. The lack of use of synthetic fertilisers and pesticides leads to restrictions in choice of crops. The loss in (present) income through a change in rotation is to some degree reflected in, and compensated by, the decrease in input costs. The success of an organic farm depends on the identification of end-uses and/or markets for all the crops in the rotation, as few farmers can afford to leave fields fallow. Wide adoption of organic agriculture is expected to change markets due to the likely prevalence of rotational crop products.

Energy: Energy in all its forms, including labour, is one of the most important agricultural inputs. Energy is required at all points of the food production chain: land preparation, planting, harvesting, transport, processing, irrigation, agro-industries and rural services. On organic farms, mechanisation is often replaced by labour, especially for weeding and harvesting in highly diversified systems. Chemical fertilisers (that are produced using a high level of non-renewable energy) are substituted with organic matter, nitrogen fixing-crops or green manuring. Indirectly, using natural rather than synthetic fertilisers saves non-renewable energy and nitrogen leaching is minimized.

Disadvantages in discarding synthetic fertiliser must be considered as well: energy needs can escalate if thermal and mechanical weeding or intensive soil tillage is used and, in some cases, organic farmers burn to clear land which reduces fertility. Many resource-poor farmers do not have access to livestock manure, often an important fertility component. Sometimes sewage sludge is used, which may contain pathogens and other contaminants. Inappropriate management of energy inputs used in organic agriculture may be detrimental to the environment: over-manuring of nitrogen has several side effects such as leaching (hence affecting water quality), stimulation of plant diseases and discouraging biological N-fixation. Environmental standards are expected to reflect local conditions so that pollution is minimized. In fact, restrictions on the number of livestock or amount of manure to be used per unit of land are not exceptional.

Diversification: Organic agriculture requires a diversity of crops and livestock. Many indigenous food crops have been supplanted by monoproduction of cash crops. Food crops, pseudocereals , grain legumes and other under-utilized plants, many of great value, can be reintroduced through crop rotations. This contributes to whole farm health, provides conservation of important genotypes, and creates habitats for beneficial species. Animal husbandry on organic farms focuses on manure production and the roughage and waste to food conversion capacity of animals.

Stability and resilience: A system resilience is its ability to overcome perturbation and to recover its function to former levels once the perturbation is removed. Resilience is of particular value to farmers in terms of risk and productivity of the system. In organic agriculture in general, a diversity of crops is grown and different kinds of livestock are kept. This diversification means that the risk in variation in production is spread, as different crops react differently to climatic variation, or have different times of growing (both in the time of the year and in length of growing period). There is therefore less chance of a bumper year for all enterprises on organic farms (likely to coincide with relatively low prices). Also, there is less chance of low production for all crops and livestock simultaneously. This contributes to food security and stability of food supply.

Yields: Factors determining yields include plant varieties and knowledge on how to manipulate biological processes within agricultural systems. The management system is a major factor in the degree of yield and financial variability. Another factor, which makes a difference in yields, is the time and length of the growth period of a crop. Exogenous factors such as climate are important.

Although yields on organic farms generally fall within an acceptable range, yield comparisons depend on the systems involved. Organic agriculture yields are lower than yields of intensive external input agriculture. The latter, however, also results in considerable losses of nutrients and environmental quality as well as in food surpluses, which are exported with subsidies. Experiences of organic production in ecosystems with low-productivity potential demonstrate the potential to double or triple average yields. The results are of course due to very low initial yields on these lands but such conditions correspond to many countries of the developing world. If similar results were to be achieved in the less endowed regions of the world, present food deficits could be partly resolved. In any case, increased yields are more likely to be achieved if the departure point is a traditional system, even if degraded, rather than a modern system.

Total farm production: It is important to consider not only yields, but also whole farm production. The total production on the farm is the yield multiplied by the area in the different crops or that used for livestock (usually measured per unit of area). When measuring production, one also needs to be aware of the concept of net production, especially relevant in developing countries. This refers to the net production of specific inputs, such as the costs of nutrients.

It should be realised that, during the conversion process, yields might be lower and investments higher than at a later stage when the organic farm has been established. The net returns to farming can therefore be lower in such a period than later.

Economic Dimensions

Farm income: The suitability of an agricultural system depends on its profitability, if that concept includes all aspects which affect farmers' welfare.

In developed countries, the financial cost of inputs (excluding labour) on organic farms can be lower than on many non-organic farms, although the magnitude differs between enterprises and countries. The difference is generally greatest in those enterprises where inputs can be readily substituted by low-cost alternatives, as fertilisers by nitrogen-fixing crops or green manure. For those inputs where substitutes are costly, such as labour cost for weeding, differences in expenditure on input between organic agriculture and other systems tend to be relatively low, or costs on organic farms can be higher. In situations where inputs are subsidised, as fertilisers and pesticides have been in a number of developing countries, the financial returns on organic farms may not be as attractive.

The legal transition to organic agriculture takes two to three years during which products cannot be sold as organic. Initial loss of yields, extent to which inputs were used under the previous management system, and the state of ecosystem degradation are often constraints that can be easy to survive only if financial support is given to farmers. Hence, the degree of support during transition, and sometimes in the first years following the transition period, are important factors in farm economics. When organic agriculture performance on the environment is rewarded (e.g. through support to conversion), organic agriculture is as profitable as conventional agriculture.

Markets: Reliable market information, quantity and regularity of supply, and comparative advantages are key to tapping market opportunities. Output prices are subject to quantity of supply and consumer willingness to pay premiums for organic products (often at prices 20 percent higher than conventional products). Entering lucrative markets entails inspection, certification and labelling of produce, the cost of which being somewhat expensive. Factors such as farm size, volume of production, and efficiency (or availability) of certification organisations determine inspection costs. Often, small farmers cannot afford certification costs, and care should be taken for not marginalising small producers.

Food security: Food security is not necessarily achieved through food self-sufficiency. Consumers' demand for organically produced food and sometimes impressive premiums provide new export opportunities for farmers of the developing world. Returns from organic agriculture have the potential, under the right circumstances, to contribute to local food security by increasing family incomes. Organic agriculture can contribute to local food security in several ways. Organic farmers do not incur high initial expenses so less money is borrowed. Synthetic inputs, unaffordable to an increasing number of resource-poor farmers due to decreased subsidies and the need for foreign currency, are not used.

Organic soil improvement may be the only economically sound system for resource-poor, small-scale farmers. This characteristic of the production process on organic farms means that organic farmers are less dependent on external inputs (e.g. fertilisers, credit), over which they may have little control, thereby increasing local food security.

Institutional Dimensions

Land tenure: Protecting soils and enhancing their fertility (land stewardship) implies ensuring productive capacity for future generations. However, in the quest to improve soil quality for the future (probably the single most important factor to determine whether farmers are interested in the issue) is whether they will benefit from the change. Security of land tenure is, therefore, an extremely important factor in this respect. If land security is not guaranteed, there is little reason for farmers to invest in a method that will bring them income in the future rather than immediate rewards.

Labour: If compared to large-scale mechanised agricultural systems, organic systems appear more labour-intensive. Using no external inputs requires that fertilisers and pest control methods be internally produced using labour intensive techniques. Labour requirements tend to be high on some organic farms (e.g. plantations), on organic farms where labour-intensive methods are used (such as strip farming, manual weeding, composting), and in low ecological potential areas that use techniques that require significant labour (such as Zai planting pits).

The timing of labour requirement throughout the year is an important aspect of labour. Care should also be paid to on-farm multiple activities (e.g. agro-tourism, processing) and off-farm labour time as often part of the family income comes from non-farm and off-farm activities. Another important issue to consider is not the quantity of labour, but the quantity of output per unit of labour, or labour productivity. While organic agriculture is likely to generate good labour productivity, labour return depends on the available amount of labour at farm household level. Often, the increase of costs of labour to produce extra units of crops is higher than the increase in outputs.

In many areas, labour is a critical constraint in agricultural production, especially when considered on a seasonal basis. Achieving high levels of labour use may not be physically possible or economically viable on small-scale farms in developing countries and may also add to household drudgery, especially for women. This is also true for cash-poor farmers and those supplementing their incomes with off-farm work. Labour and total costs are generally lower on private organic farms.

In the developed world, labour scarcity and costs may deter farmers from adopting organic management systems. However, where labour is not such a constraint, organic agriculture can provide employment opportunities in rural communities. Furthermore, the diversification of crops typically found on organic farms, with their various planting and harvesting schedules, may result in more work opportunities for women and a more evenly distributed labour demand which helps stabilise local/ regional employment.

Justice: The concept of "fair trade" implies a concern of the buyer for social justice for those who work in agriculture, especially with regard to a "fair wage". In fair-trade projects, traders ensure that producers receive a minimum return for their produce irrespective of the actual market price. At present, certification that guarantees fair trade does not necessarily imply organic production. Organic certification organisations favourably consider inclusion of "reasonable wage conditions".

Improving the situation of women is recognised as an important issue within organic agriculture. The social environment of the workers engaged in organic agriculture, namely their working conditions and environment, often results in improved housing situations and childcare facilities. Availability of work, gender distribution of labour and access to knowledge are key considerations in organic agriculture.

Attitudes and perceptions: The single biggest constraint to the development of organic agriculture is that most people in all kinds of areas (including scientists, researchers, extension officers and politicians) strongly believe that organic agriculture is not a feasible option to improve food security. For this reason, very few farmers can obtain information about this management system, even when they inquire about it. If those who make policy decisions on the allocation of resources, such as for research and extension, are not aware of the possibilities of organic agriculture, no positive consideration towards this farming system can be expected.

Investments and benefits: The advancement of organic agriculture to date has to a large extent been due to private investment. This has been in the form of consumers' willingness to pay for organic commodities (price premiums) and farmers' readiness to experiment and innovate, despite the risks involved with such on-farm research. Many of the inputs used in organic agriculture are public goods (such as knowledge about practices) which can be used without impeding use by others. Hence, there is little private interest in promoting particular inputs, which are used in organic agriculture.

Knowledge: Empirical knowledge of natural processes takes a long time to consolidate itself. Traditional knowledge can be improved through selective introduction of results of modern science in areas such as energy flows and bio-geochemical cycles, biotic and abiotic factors, which regulate plant development, renewable energy technologies, and management techniques. The integration of traditional technologies with emerging technologies would lead to an improvement of traditional technologies in terms of: i) cost per unit; ii) productivity per unit of factor input; and iii) quantity and quality of output.

Community participation: Engaging in organic production means experimenting new techniques, introducing different management of labour time, investing efforts in different management of space, adapting and refining solutions to change, comparing different options with farmers that have similar conditions, and making appropriate choices. This can only be achieved through farmers' participation in research and its application.

This on-farm research component can support rural communities, and generate new knowledge that will benefit all farmers. Consistent labour needs, combined with the enhanced capacity of the land and protection of water associated with organic agriculture, may encourage people to permanently locate and thus reinvigorate rural communities. The establishment of cooperation between farmers is instrumental in helping farmers to become a stronger and more independent partner in the agro-business environment.

Self-reliance: Within organic agriculture, the use of locally available inputs is encouraged. The effect on the local community of such a form of agriculture is, therefore, likely to be greater than when inputs are imported from outside the community. In those cases where synthetic fertilisers and pesticides are imported, adoption of organic agriculture techniques means a decrease in imports, decreasing the need for foreign currency and credit.

The site-specific nature of organic agriculture also means that indigenous species and knowledge, so often discounted, are of great value. In many places, this knowledge has been eroded with the introduction of high external input agriculture, promotion of monoculture, and selection of "improved products." Farmers may readily welcome a management system close to their own traditions and not driven solely by a production ethic.

Policies and regulations: Norms, standards, inspection and certification in organic agriculture have been established and remain enacted by the private sector. Recently, a number of countries have established guidelines, regulations and policies to assist producers to respond to the growing demand for organic food and to control fraudulent practices.

Favourable policies and institutional structures are important incentives to the adoption of organic agriculture. Where supportive government policies are lacking, and especially where inspection organisations are unavailable (or need to be imported at high expense), interested communities need to mobilise and organise themselves to procure necessary inputs and to access markets.

Socio-Ecological Analysis

Market food prices frequently fail to reflect the true cost of production, as they do not internalise environmental and societal costs of production. The pursuit of short-sighted economic gain, coupled to the lack of awareness of the long-term implications of unsustainable agricultural practices undermines the basis of agricultural production. Management in time and space of interactions between ecological, economic and social variability is lacking for sound decision-making.

Governments, farmers and consumers make their decisions according to the information available to them. Sound information on the status of renewable resources, natural resources dynamics, land-use and tenure patterns, sociocultural knowledge and conditions are necessary to chose among different agricultural production systems and invest resources such as capital, labour, markets structures and support services.

Agro-ecosystems Information Collection

Biophysical and socioeconomic parameters allow the understanding of social dynamic of farming and resource-use systems. Information collection of agro-ecological, economic, social and institutional parameters needs to cover:

i) the state of resources;

ii) resource management efficiency and where possible,

iii) anticipated opportunities and constraints.

Agro-ecological parameters

Resources

— *land:* availability of land and soil quality and productivity ;

— *water:* water quality and hydrology (water supply, hydric balance and soil occupation);

— *biodiversity:* ecosystem and species diversity, distribution and abundance;

— *ecosystems:* habitats and ecological relationships that determine productivity, presence of special ecosystem or species, biomass, primary and secondary productivity, eco-regulation (degree of pest and disease occurrence, pest predator presence), carrying capacity of the system (e.g. stocking rate);

— *climate:* winds, rainfall, and climate variability.

Resource management efficiency

— flows of organic matter and nutrients: nutrient balance (excess, depletion), quantity of organic matter, organic matter recycled internally, ratio of internal and external nutrient inputs, macro- and micro-nutrients in soil, ratio of on-farm feed production and imported feed;

— *water:* water quality and water balance;

— *biodiversity:* number and quantity of crop and animal varieties and species, crop rotation index, composition of micro- and macro-soil life, pest incidence, resistance in genetic resources and needs for pesticide use;

— *productivity:* genetic potential of crops/breeds to yield, growth and yields ;

— *climate:* creation of micro-climates, emissions from livestock;

— *energy:* ratio of renewable and fossil energy used, ratio of internal and external energy inputs used, and ratio of labour substitution to mechanisation.

Anticipated opportunities and constraints

— crop protection strategy: which pests are common,

what is the cause, what can be done to avoid them within available resources;

— access to material which can be used in compost and mulch, availability of other materials such as rock dust;

— import needs: organic fertilisers and pesticides, seed/breed, water, energy;

— pollution and degradation to be mitigated and means available;

— long-term productivity: the effect of present production on the soil and biodiversity and implications for future yields;

— added valued: contribution of livestock to the farm, benefits from trees and biomass, resilience.

Economic parameters

Resources

— availability of capital, degree of financial autonomy and level of investment;

— financial incentives: government economic support during and/or after conversion, preferential conditions of credit, tax reduction;

— existing storage, handling, processing and transportation facilities for organic produce;

— cost of inspection and certification to total production;

— market opportunities: domestic and international markets, level of price premium, volume, quality and regularity of supply, comparative advantage.

Resource management efficiency

— use of production factors: value of produce compared

to production costs, production per hectare, per unit of water, per farm worker, per unit of capital;

— number of on- and off-farm activities that contribute to farm income;

— ratio of the value of external inputs and the total production value;

— investment in enterprise development (e.g. equipment and infrastructures).

Anticipated opportunities and constraints

— assessment of economic variability: risks and variability of yields in good and bad years, market price change, government intervention, exchange rates;

— evaluation of non-natural resources needed, such as capital, machinery and tools;

— total production of all enterprises (not only of the main enterprise);

— total net return: income from main crop and other crops and livestock, minus the cost of the inputs used for the production;

— return on capital: a balance between additional return (including also reduction in costs) and extra costs (including also revenue foregone).

Social parameters

Resources

— human assets in term of skills (and training needs);

— labour requirements (quantity and timing of labour) throughout the year;

— people expected to derive livelihoods (income) from the enterprise;

— land tenure and conditions of resources access;

— gender desegregated access to resources and benefits;

— income distribution and gender division of labour.

Resource management efficiency

— labour intensity per hectare and unit of produce and labour productivity;

— total food availability per family compared to consumption requirements.

Anticipated opportunities and constraints

— degree of risk;

— rural employment and migration patterns;

— household food security and nutrition;

— degree of self-reliance.

Institutional parameters

Resources

— existing laws, standards for organic production, processing, inspection and labeling;

— availability of government support: research, technical assistance, training and support to marketing;

— availability of inspectors and certifiers (domestic or international);

— contacts with organic processors, distributors, traders and retailers;

— marketing outlets: farm gates, home delivery, specialised shops, supermarkets, and export markets.

Resource management efficiency

— community structures (e.g. farmers' associations) for self-certification;

- — farmers cooperation and exchange of information;
- — on-farm research: degree of farmers participation in experiments and percentage of farmers adopting and replicating experiments;
- — number of farmer representatives in government discussion fora.

Anticipated opportunities and constraints

- — difficulty to comply with rules and standards;
- — insufficient access to support services;
- — good governance;
- — social cohesion.

Availability of natural and human resources differ from location to location. Hence, the practicality of certain practices, their opportunities and constraints, will vary considerably between localities. It is not sufficient to consider the variables within a single dimension. In fact, the performance of each of the agro-ecological, economic, social and institutional dimensions involves trade-offs among them.

Management options should not only aim at optimising agronomic performance but also at optimising the ecological and social performance of the system. For example, maintaining soil fertility offers the choice between applying compost or green manuring: if organic material and labour are available, the first option is the one that can perform better on the agro-ecological, economic and social fronts.

Different priorities express different attitudes towards nature, society and the ethical decisions involved. In line with these differences, different values (expressed as priorities) will have specific consequences on equity, economy and environmental sustainability. Farming objectives such as environmental and landscape restoration, rural development and local food security or export markets orientation will entail different priorities and value systems.

Specific farming objectives have to be set in light of identified needs, opportunities and constraints. It is with a specific objective in mind that farmers decide on what to produce, how much to produce, and how best to produce. Objectives and constraints of stakeholders at the micro-level (e.g. farmer or community) should be integrate(e.g. policy, markets) and different types of trade-offs evaluated.

4

Organic Agriculture for Sustainable Food Production

Organic farming is one of the fastest growing segments of agriculture during the 1990s. USDA estimates that the value of retail sales of organic foods in 1999 was approximately $6 billion. The number of organic farmers is increasing by about 12 percent per year and now stands at about 12,200 nationwide, most of them small-scale producers.

The Organic Foods Production Act (OFPA), enacted under Title 21 of the 1990 Farm Bill, served to establish uniform national standards for the production and handling of foods labelled as "organic." The Act authorised a new USDA National Organic Programme (NOP) to set national standards for the production, handling, and processing of organically grown agricultural products. In addition, the Programme oversees mandatory certification of organic production.

The Act also established the National Organic Standards Board (NOSB) which advised the Secretary of Agriculture in setting the standards upon which the NOP is based. Producers who meet standards set by the NOP may label their products as "USDA Certified Organic."

"The new organic standard offers a national definition for the term "organic." It details the methods, practices and substances that can be used in producing and handling organic crops and livestock, as well as processed products. It establishes clear organic labelling criteria, and specifically prohibits the use of genetic engineering methods, ionising radiation, and sewage sludge for fertilisation."

Problems

Environmental degradation: After half a century of high intensive input agriculture, the yield gap between best practices and farmers' fields remains large, agricultural lands continue to shrink and global environmental threats are a reality, e.g. erosion of biodiversity, desertification, climate change and other transboundary pollution. Agriculture intensification contributes to over 20 percent of global anthropogenic greenhouse gas emissions. Agricultural activities affect 70 percent of all threatened bird species and 49 percent of all plant species.

Uniform cultures have dramatically reduced the number of plants and animals used in agriculture; currently, 1 350 breeds face extinction, with two breeds being lost each week. Biodiversity erosion is exacerbated by the loss of forest cover, coastal wetlands and other wild relatives, important for the development of biodiversity and essential for food provision, particularly in times of crisis.

Hunger: Despite FAO's and other institutions' efforts for global food security, the benefits of increased agricultural production often bypass the poorest sections of the world population. Over 450 million farmers have never had access to Green Revolution technologies such as mechanisation, irrigation, improved seeds and breeds, and synthetic fertilisers and pesticides. The World Food Summit stressed that hunger is a problem of access to means of food production or means to purchase food.

Dependence: Although agriculture remains the world's single largest employer, rural economies suffer from sharp decline in real food prices and loss of entrepreneurial capacity. The terms of international exchange favour importations at the expense of local production. Developing countries, which traditionally have had a net surplus in agricultural trade, increasingly depend on food imports. The conventional food production model ties farmers into conditions of dependence on large corporations to buy agricultural inputs (seeds, fertilisers, pesticides) and to sell their produce.

Paradigm Shift

Nature conservation: Protecting biodiversity at genes, species and ecosystem levels through germplasm banks and protected areas is not sufficient. The maintenance costs of gene banks are high, up to half of the material collected is in need of regeneration, and "freesing" genetic resources denies their evolution. Biodiversity is best maintained through sustainable utilisation and selection by food providers.

Animals move across boundaries and ecosystems are not immune to air and water pollution. The 12 percent of global land areas "fenced" for nature protection are located within or around a 40 percent land surface used by agriculture and forestry. Food systems should be viewed as an integral part of the ecosystem. "There is a need to manage agricultural land as part of a larger landscape that explicitly considers ecological functioning".

Rural livelihoods: Rural communities need land to derive livelihoods. National parks and protected areas have always had local culture as components of the landscape. Ecotourism management by local residents generates supplementary income for both management of protected areas and rural livelihoods. Agritourism creates new opportunities for farm employment, specialty foods and safe agro-ecosystems. There is a need to reconnect farmers to the land by linking their productive activities with nature conservation. A symbiotic relationship between agriculture and natural landscape exists in the presence of ecologically managed systems.

Food self-reliance: Poor farmers and market-marginalised communities need a food production model that relies on local natural resources and ecological management. Where food surpluses are generated, markets should guarantee "fair" prices. Managing local resources without having to rely on external inputs involves substituting purchased (private) goods by (public) knowledge of natural processes that optimise competition for nutrients and space between species within the agro-ecosystem.

By stressing diversification and adaptive management, agricultural systems can improve soil and water quality and the ecological services that support agriculture, while significantly decreasing vulnerability to weather vagaries or other factors. Where markets for ecologically-produced food exist, farmers' return on labour are higher through consumers' willingness to pay higher prices.

Biodiversity in International Agreements

Organic agriculture offers a means to address food self-reliance, rural development and nature conservation. The common thread in this ambitious goal is the sustainable use of biodiversity; in terms of both agriculture contribution to biodiversity and biodiversity contribution to agriculture.

To be successful, organic agriculture needs functional groups of species and essential ecosystem processes as its main "input" to compensate for the restriction on (or lack of) synthetic input use. In fact, a close relationship exists between organic agriculture and the maintenance of biodiversity. This is expressed in the rules and regulations that govern certified organic agriculture and by the practical experiences of organic farmers around the world. Selected international codes and agreements expressing the organic agriculture-biodiversity linkages and need for further consideration are the following:

— The Vignola Declaration and Action Plan on the Relationship Between Nature Conservation, Biodiversity and Organic Agriculture of the Intern-ational Federation for Organic Agriculture Movements, the World Conservation Union and the World Wide Fund for Nature. This Declaration recognises that "organic agriculture is essential for conserving biodiversity and Nature" and "asks environmentalists, farmers, politicians, industry and international institutions to support and develop organic agriculture as the most ecologically-sound agricultural system".

— The FAO/WHO Codex Alimentarius guidelines for Organic Food define organic agriculture as "a holistic production management system which promotes and enhances agro-ecosystem health, including biodiversity, biological cycles, and soil biological activity. The primary goal of organic agriculture is to optimise the health and productivity of interdependent communities of soil life, plants, animals and people".

— The Convention on Biological Diversity (CBD) "encourages the development of technologies and farming practices that not only increase productivity, but also arrest degradation as well as reclaim, rehabilitate, restore and enhance biological diversity and monitor adverse effects on sustainable agricultural biodiversity" such as "inter alia, organic farming". Furthermore, the CBD Global Strategy for Plant Conservation includes outcome-oriented global targets for 2010 whereby using plant diversity sustainably considers, among others, organic food as an indicator of direct measures to reach the target of "30 percent of plant-based products derived from sources that are sustainably managed".

— The Ninth Session of the Commission on Genetic Resources for Food and Agriculture recommended that FAO further studies the "contribution of genetic resources to the economies of member countries, the impact of agricultural subsidies on biodiversity, and the competitiveness of organic agriculture".

Role of Organic Agriculture in Enhancing Biodiversity

In organic agriculture, biodiversity is both instrument and aim. Natural ecological balance, below and above ground, is key to its success. A healthy soil is the base for food production and a diversity of plants and animals on land prevents pest and disease outbreaks. Although organic agriculture is committed to the conservation and enhancement of biodiversity, many systems today remain limited to input substitution. To be unlocked, the real potential of organic agriculture on biodiversity requires a stronger shift to a systems approach, based on improved understanding of ecosystem functions.

Soil Biodiversity

Living Organisms

Soils contain enormous numbers of diverse living

organisms assembled in complex and varied communities. Soil biodiversity reflects the variability among living organisms in the soil—ranging from the myriad of invisible microbes, bacteria and fungi to the more familiar macro-fauna such as earthworms and termites. Plant roots can also be considered as soil organisms in view of their symbiotic relationships and interactions with other soil components.

These diverse organisms interact with one another and with the various plants and animals in the ecosystem, forming a complex web of biological activity. Environmental factors, such as temperature, moisture and acidity, as well as anthropogenic actions, in particular, agricultural and forestry management practices, affect to different extents soil biological communities and their functions.

Soil organisms contribute a wide range of essential services to the sustainable functioning of all ecosystems. They act as the primary driving agents of: nutrient cycling, regulating the dynamics of soil organic matter, soil carbon sequestration and greenhouse gas emissions; modifying soil physical structure and water regimes; enhancing the amount and efficiency of nutrient acquisition by the vegetation; and enhancing plant health. These services are not only critical to the functioning of natural ecosystems but constitute an important resource for sustainable agricultural systems.

Healthy Soils

Capturing the benefits of soil biological activity for agricultural production requires adhering to the following ecological principles:

— *Supply of organic matter:* Each type of soil organism occupies a different niche in the web of life and favours a different substrate and nutrient source. Most soil organisms rely on organic matter for food; thus a rich supply and varied source of organic matter will generally support a wider variety of organisms.

— *Increase of plant varieties:* Crops should be mixed and their spatial-temporal distribution varied, to create a greater diversity of niches and resources that stimulate soil biodiversity.

— *Protection of soil organisms' habitats:* The activity of soil biodiversity can be stimulated by improving soil living conditions, such as aeration, temperature, moisture, and nutrient quantity and quality. In this regard, reduced soil tillage and minimized compaction—and refraining from chemical use—are of particular note.

Improvement in agricultural sustainability requires, alongside effective water and crop management, the optimal use and management of soil fertility and soil physical properties. Both rely on soil biological processes and soil biodiversity. This calls for the widespread adoption of management practices that enhance soil biological activity and thereby build up long-term soil productivity and health.

Adaptation and further development of soil biodiversity management into sustainable land management practices requires solutions that pay adequate consideration to the synergies between the soil ecosystem and its productive capacity and agro-ecosystem health.

Soil Fertilisation

Building of soil fertility is the cornerstone of organic agriculture. Organic practices create suitable conditions for soil biotic and abiotic resources through: manipulation of crop rotations and strip-cropping; green manuring and organic fertilisation (animal manure, compost, crop residues); minimum tillage; and avoidance of pesticides and herbicides use. Scientific research in Europe has demonstrated that organically-managed soils significantly increase biological activity and total density and diversity of soil micro-organisms. Such biodiversity enhances nutrients recycling and soil structure. The impact of organic management on soil biological activity and related benefits is summarised below:

— *Abundant earthworms and arthropods:* Organic management increases the abundance and species richness of earthworms and beneficial arthropods living above ground, and thus improves the growth conditions of crops. The biomass of earthworms in organic systems is 30-40 percent higher than in conventional systems, their density even 50-80 percent higher. Compared to the mineral fertiliser system, this difference is even more pronounced. More abundant predators help to control harmful organisms (i.e. pests). In organic systems, the density and abundance of arthropods, as compared to conventional systems, has up to 100 percent more carabids, 60-70 percent more staphylinids and 70-120 percent more spiders. This difference is explained by prey deficiency due to pesticide influence as well as by a richer weed flora in the standing crop that is less dense than in conventional plots. In the presence of field margins and hedges, beneficial arthropods are further enhanced, as these habitats are essential for over-wintering and hibernation.

— *High occurrence of symbionts:* Organic crops profit from root symbioses and are better able to exploit the soil. On average, mycorrhisal colonisation of roots is highest in crops of unfertilised systems, followed by organic systems. Conventional crops have colonisation levels that are 30 percent lower. The most intense mycorrhisal root colonisation is found in grass-clover, followed by the vetch rye intercrop. Roots of winter wheat are scarcely colonised. Even when all soils are inoculated with active micorrhisae, colonisation is enhanced in organic soil. This indicates that, even at an inoculum in surplus, soil nutrients at elevated levels and plant protection suppress symbiosis. This underlines the importance of appropriate living conditions for specific organisms.

— *High occurrence of microorganisms:* Earthworms work hand-in-hand with fungi, bacteria, and numerous other microorganisms in soil. In organically managed soils, the activity of these organisms is higher. Microorganisms in organic soils not only mineralise more actively, but also contribute to the build up of stable soil organic matter. The amount of microbial biomass and decomposition is connected: at high microbial biomass levels, little light fraction material remains undercomposed and vice versa. Thus, nutrients are recycled faster and soil structure is improved.

— *Microbial carbon:* The total mass of microorganisms in organic systems is 20-40 percent higher than in the conventional system with manure and 60-85 percent higher than in the conventional system without manure. The ratio of microbial carbon to total soil organic carbon is higher in organic systems as compared to conventional systems. The difference is significant at 60 cm depth (at 80 cm depth, no difference is observed). Organic management promotes microbial carbon and thus, soil carbon sequestration potential.

— *Enzymes:* Microbes have activities with important functions in the soil system: soil enzymes indicate these functions. The total activity of microorganisms can be estimated by measuring the activity of a living cell-associated enzyme such as dehydrogenase. This enzyme plays a major role in the respiratory pathway. Proteases in soil, where most organic N is protein, cleave protein compounds. Phosphatases cleave organic phosphorus compounds and thus provide a link between the plant and the stock of organic phosphorus in the soil. Enzyme activity in organic soils is markedly higher than in conventional soils. Microbial biomass and enzyme activities are closely related to soil acidity and soil organic matter content.

— *Wild flora:* Large organic fields (over 15 ha) feature flora as six times more abundant than conventional fields, including endangered varieties. In organic grassland, the average number of herb species was found to be 25 percent more than in conventional grassland, including some species in decline. Vegetation structure and plant communities in organic grassland are more even and more typical for a specific site than in conventionally managed systems. In particular, field margin strips of organic farms and semi-natural habitats conserve weed species listed as endangered or at risk of extinction. Animal grasing behaviour or routing activity (e.g. pigs) was found important in enhancing plant species composition. Weeds (often sown in strips in organic orchards to reduce the incidence of aphids) influence the diversity and abundance of arthropods and flowering weeds are particularly beneficial to pollinators and parasitoids.

— *High-energy efficiency:* Organic agriculture follows the ecosystem theory of a closed (or semi-closed) nutrient cycle on the farm. Organic land management allows the development of a relatively rich weed-flora as compared to conventional systems. Some "accompanying plants" of a crop are considered useful in organic management. The presence of versatile flora attracts beneficial herbivores and other organisms that improve the nourishment of predatory arthropods. When comparing diversity and the demand of energy for microbial maintenance, it becomes evident that diverse population need less energy per unit biomass. A diverse microbial population, as present in the organic field plots, diverts a greater part of the available carbon to microbial growth. This increases the turnover of organic matter, with a faster mineralisation and delivery of plant nutrients. Finally, more organic matter is diverted to build up stable soil humus.

— *Erosion control:* Organic soil management improves soil structure by increasing soil activity and thus, reduces erosion risk. Organic matter has a positive effect on the development and stability of soil structure. Silty and loamy soils profit from organic matter by an enhanced aggregate structure. Organic matter is adsorbed to the charged surfaces of clay minerals. The negative charge decreases with increasing particle size. Silt is very susceptible to erosion since it is not charged, but organic-matter layers on the silt surface also favour aggregates with silt.

Genetic Resources for Food and Agriculture

There are several hundred millions of small farmers in the world who do not have the economic means to buy high yielding seeds or the synthetic fertilisers and pesticides necessary for conventional cultivation. Many of these have opted for the maintenance or reintroduction of organic systems based on traditional forms of agriculture. These promote the use of varieties and breeds that are better adapted to local stress conditions and do not require unavailable inputs such as, for example, veterinary drugs.

There are also farmers who have opted for organic agriculture, in part because they wish to produce healthy and environmentally-friendly food, and also because they are attracted by the strong demand for organic products and the related premium prices. Market driven farmers should, as a minimum, rotate crops as the first step towards improving agricultural biodiversity. This is one of the methods required by organic certification bodies as well as by financial programmes. These farmers have also opted for sowing locally-adapted species and varieties that are more resistant to disease and local environmental conditions because synthetic fertilisers and pesticides cannot be relied upon.

The adoption of organic agriculture methods requires farmers to follow a series of agronomic practices that make organically managed systems biologically much more complex than conventionally managed systems. Organic farms make use of larger numbers of plant and animal species than conventional systems. As a result, the large pool of genetic resources for food is maintained and other useful organisms, such as predators, pollinators and soil microorganisms are increased—for the very benefit of the agricultural system.

Centres of Diversity

The continued cultivation of crop species within their centres of diversity plays a fundamental role in the maintenance of genetic diversity. Preserving the integrity of centres of diversity through ecologically-sound agriculture is an indispensable inheritance for agriculture and as such, for food security for future generations. It is the genetic variability that allows population to adapt to changing environmental conditions.

In centres of diversity, the introduction of organic practices is aiding the conservation, through cultivation, of population with high genetic variability. The maintenance of agricultural production in centres of diversity requires market outlets for peasants and indigenous communities. The new income opportunities offered by organic markets reverse the present trend of abandoning land which has previously been economically uncompetitive.

Examples where organic agriculture created viable means for in situ conservation and use of areas with diverse genetic heritage include: producing and processing cocoa in Mexico to sell Maya chocolate to tourists, and the maintenance of naturally-pigmented cotton in Peru, resistant to pests, diseases and drought. These organic market opportunities provide for the economic viability of thousands of farmers and processors and maintain genetic variability for future generations.

Revival of Under-utilised Species, Varieties and Breeds

In the past, agriculture has played an important role in the maintenance of genetic diversity. The substitution of a large quantity of species for only a few and the adoption of high yielding and uniform varieties from a genetic point of view, has caused a significant reduction in the genetic inheritance of cultivated species. Many agricultural species, varieties and breeds which have played an important role in the human diet and traditional cultures have practically disappeared over the last century.

Organic farmers breed varieties for quality, nutrition, resistance and yield, in reduced input growing conditions. Research has shown that these characteristics are more likely to be found in older native cultivars. In particular, open pollinated varieties and indigenous breeds offer diverse and regionally adapted characteristics that are better suited to organic agriculture.

In the last decade, the adoption of organic agriculture has indirectly established a rescue process of species, varieties and breeds threatened by under-use or extinction. The restoration and enhancement of under-utilised species and varieties has been motivated by specific demands of both consumers and farmers.

For the rescue of varieties threatened by extinction, the development of a market is fundamental and it is here that organic agriculture plays an important role as the price premium gives an additional value to the product. This is especially the case now that there is a consumer' interest in traditional, speciality and organic products.

The majority of crop varieties available on the commercial market are not suitable for organic cultivation methods as they have been selected for production dependent on irrigation and large quantities of synthetic fertilisers and pesticides. Many of these are hybrids and are not open-pollinated. In the last few years, the problem has worsened following the arrival on the market of genetically modified varieties.

The selection objectives of organic agriculture differ from those for conventional agriculture. It is of crucial importance to utilise the genotype potential for an improved adaptation of varieties to the low-input conditions prevailing in organic agriculture.

The necessity for organic farmers to find methods for obtaining quality products with good yields and limited production costs is greater than for other farmers. Besides the fact that organic farmers cannot apply synthetic inputs, their use of organic fertilisers, natural pesticides and other permitted substances is uneconomical in the long-term.

Permitted external inputs as such are relied upon mainly during the conversion period to organic agriculture or under exceptional circumstances. The comparative advantage of certain varieties to withstand local natural stress, especially in marginal areas, leads organic farmers to adopt biodiversity management as their main productive strategy.

Empirical organic breeding systems have been based on the selection of individuals better adapted to the local environment and that are more resistant to pests and diseases. Many of these systems have demonstrated interesting results in restoring and improving local varieties.

Animal breeding for high performance and selecting for early maturity have led to increased susceptibility to diseases, joint inflammation and mastitis as well as circulatory, metabolic and fertility problems of livestock. Loss of breeds is exacerbated by the narrowing genetic base of modern breeds and hybrid lines. The trend towards inbreeding increases the degree of genetic uniformity in the animals and hence, susceptibility to infection, parasites or epidemics.

A significant proportion of local breeds remains in the care of pastoral people and traditional livestock owners in developing countries (e.g. pigs in China, cows in India and poultry in Asia and Latin America). Local breeds are suitable for free ranging and robust, thus viable in marginal areas. While the yield may be less in the short-term, animals are more resilient and able to survive in the long-run.

Distribution of Organic Genetic Resources

Historically, farmers have managed many varieties and breeds according to agronomic and culinary properties.

Considering the need for a wide gene pool to improve and multiply genetic resources for food and agriculture, breeding requires access to seeds and breeds from the formal and informal sectors. Open pollinated varieties, which represent an important gene pool for resource-poor farmers living in marginalised and stress-prone areas, are rapidly vanishing. They are replaced by very few hybrid varieties which require inputs not available to poor farmers and which entail dependence on large seed companies.

Limitations and threats associated with crops have stimulated many organic farmers, especially in the horticulture sector, to produce their own seeds. In order to do this, they have often had to rescue local varieties and develop their own system of selection and distribution. In many cases, the systems include the exchange of seeds between farmers as a fundamental instrument (e.g. organic seed fairs).

Organic systems encourage the preservation and expansion of older, locally bred and indigenous varieties and breeds. Farmers who save their own seeds can gradually increase crop resistance to pests and diseases by breeding for "horizontal resistance". Horizontal resistance is the ability of a crop to resist many or all strains of a particular pest (which differs from breeding for "vertical resistance" to have a gene to resist one specific strain of a disease).

By exposing a population of plants to a certain disease or pest (or to several pests at one time), then selecting a group of the most resistant plants and interbreeding them for several generations, a given population becomes more resistant than the original population. Horizontally resistant cultivars are well adapted to the environment in which they were bred, but may be less suitable for other growing conditions.

Benefits derived from new varieties bred by farmers require a legal system of common ownership that allows equitable access and benefit sharing. The biodynamic network of farmers and breeders in Germany provides an example of how such a system could be organized: trials, selection and evaluation of genotypes adapted to low-input conditions is made by farmers and common ownership of new varieties is shared by the community.

Organic agriculture is providing an important contribution to the in situ conservation, restoration and maintenance of agricultural biodiversity. The spontaneous establishment of participatory systems of research and development is shaping a simple and practical system of equitable sharing of benefits derived from genetic resources for food and agriculture. The growth pattern shown by the conversion to organic agriculture throughout the world suggests that this contribution is likely to increase still further.

Wildlife Biodiversity

Wildlife and Agriculture

Agricultural productivity depends upon the maintenance of ecological balances and the natural properties of plants and animals. The fundamental role of maintaining (or restoring) biodiversity is demonstrated through ecological services such as pollination of crops, predation for biological control of pests, microorganisms' maintenance of soil fertility and other services vital to the food web.

On the other hand, agriculture has the same important role in wildlife conservation, provided that it avoids the use of substances that could have a harmful effect on natural species and that it maintains food and shelter through a diversified landscape and permanent vegetation. Finally, a type of land use that provides suitable biological corridors is essential for wildlife conservation.

Nature conservation has traditionally consisted of geographically targeted efforts. While this approach has resulted in a number of successes for rare species or key locations, worrying declines of protected species have occurred. A healthy environment is a prime objective for the conservation of vital terrestrial ecosystems and the wildlife in it. Natural faunal and floral species have strong connections with agriculture, whatever their habitats are, especially as agricultural fields occupy much of the earth's land surface.

Protected areas simply cannot be viewed in isolation from the communities within and near them. Almost everywhere, rural dwellers claim historical rights on protected areas which governments have, at a point in time, declared "protected" for national interest. People inhabiting within or in the neighbourhood of protected areas depend directly on their resources for a living.

Nature Conservation

There is no doubt that farmers are the most important managers of natural resources. Several studies indicate that organic agriculture safeguards non-agricultural biodiversity and offers a viable alternative in protected area categories where human activities are allowed. Most importantly, the huge land surface surrounding protected areas requires an agro-ecosystem management that preserves the safety and integrity of the landscape. If farm land bordering and connecting protected areas employ organic methods, there is no reason to fear the loss of wildlife or contamination of air, water and soil. These buffer zones are critical to the success of conservation in the protected areas.

Organic agriculture enhances people's ability to live in harmony with nature and to derive economic benefit from their land. Considering that most protected areas traditionally belonged to local villagers, organic agriculture allows local people to maintain some control over their land, protect land and biodiversity through their farming practices, reap its benefits for themselves and, at the same time, conserve and improve the natural environment.

Protected Areas and Buffer Zones

Certain protected area categories allow sustainable land-use activities such as organic agriculture, management of non-timber forest products, fishing, subsistence hunting and ecotourism. Organic farming within protected areas is a growing practice that can "help to define and control sustainable land uses in those protected areas containing significant human population".

There are some 350 Biosphere Reserves in 85 countries to protect ecosystems. The area surrounding a biosphere reserve, named buffer zone, plays a critical role because activities carried out here strongly influence the core of the protected area itself. Often buffer zones are areas dedicated to agricultural practices. Conversion to organic systems can reduce the detrimental effects of conventional farming, and can provide sustainable systems, suitable for the management restrictions governing buffer zones, and consequently to natural ecosystem conservation.

The protection of the natural heritage must consider the impact, be it positive or negative, that human activities have on it. The characteristics of agriculture make it one of the main activities to be practised in protected areas and buffer zones. Organic agriculture offers a suitable alternative in ecosystems where geographic and morphological conditions are favourable to human activities, such as wetlands and lowland forests. Also, the ecological services offered by organic agriculture in biological corridors are of extreme importance.

Wetlands are defined in the Ramsar Convention as "areas where water is the primary factor controlling the environment and the associated plant and animal life. They occur where the water table is at or near the surface of the land, or where the land is covered by shallow water". Wetlands are present in every country, from the tundra to the tropics. Their ecological importance derives from their capacity to host high concentrations of birds, mammals, reptiles, amphibians, fish and invertebrate species. In fact, wetlands are boundary areas that combine the components of marine, fluvial and terrestrial ecosystems.

In addition, many functions of wetlands derive from the interactions of different properties of soils, water, plants and animals. Some of these functions include, for example, water purification, water storage, flood mitigation, recharge and discharge of underground aquifers (by the movement of water), and stabilisation of local climatic conditions.

Natural wetlands are among the most threatened ecosystems in the world. Their high productivity and the morphologic characteristics make wetlands excellent areas for many human activities, especially agriculture. Often the damage caused by land reclamation for agriculture or unsustainable practices leads to the disappearance of wetland areas and corresponding biodiversity. Suitable management is therefore a critical priority to save these fragile ecosystems.

Organic agriculture can help wetland conservation, by providing suitable habitats for wildlife species, reducing water pollution and, at the same time, offering a valid economic alternative to the exploitation of natural resources.

Human activities such as modern agriculture and grasing can be a serious threat to forest ecosystems and hence, for a large percentage of the world's flora and fauna. In fact, the practice of clearing the tree cover often interrupts the continuity of canopy, an essential characteristic for genetic and specific biodiversity flow. In many cases, the same agricultural fields are also a barrier to wildlife movement, and pollution from agrochemical abuse may have negative impacts on forests.

Organic management in forest areas has the potential to reduce the loss of biodiversity caused by these agricultural activities. The organic systems utilised in forest areas are shade and sun cultivations. In the first case, the plants are grown under a canopy that consists of original forest trees or selected plants. In the latter, there is no forest cover on the agricultural field.

Although yields of full sun production are often higher than for shade cultivations, the latter provide timber, fuelwood and other fruits, and have a lower soil erosion rate. Shade plantations have been shown to be highly beneficial to biodiversity conservation in tropical forest ecosystems, including millions of migratory birds and other animals and plants.

The Northwest Shade Coffee Campaign has obtained statistical data on shade coffee as compared to sun coffee and has shown how shade cultivations are preferred by bird species (approximately 150 compared with 20-50 species), mid-size mammals (24 species compared with almost none) and several species of ants, beetles, amphibians, epiphytes and others. Smithsonian Migratory Bird Centre's studies indicates that at least 180 species of birds live in Mexican shade coffee and cocoa fields (much more than on other agricultural lands), and that 90 percent fewer bird species live in Colombian sun coffee plantations than in shade coffee.

Researchers highlighted the capacity of the canopy cover to support secondary structures (e.g. epiphytes, parasites, mosses and lichens), which in turn support arthropods, amphibians and other living beings. Another relevant fact is that shade coffee hosts a large density and diversity of predators and parasitoids involved in the control of insect pests. Shade coffee offers an optimal habitat also for other tropical forest species, such as beetles, ants, wasps and spiders, and supports a high diversity of many vertebrate groups (small mammals as opossums, squirrels, mice and bats). Finally, areas of high ecological value located around coffee farms are protected.

Corridor zones link protected areas with one another and either remains under wild cover, or are managed to ensure that human land use is compatible with the maintenance of a high degree of biological connectivity. These areas have a fundamental ecosystem function forming part of a large-scale ecological web.

To be effective, biological corridors must offer suitable habitats to wildlife, thus the maintenance of a healthy environment is necessary. Areas joining parks or reserves are equally important for biodiversity conservation as they ensure ecosystem connectivity. These "linkage" areas often host agricultural systems but if they are managed in an unsustainable manner, the vitality of these corridors decreases or is totally hindered.

Careful organic management in agricultural land between (and sometimes inside) the protected areas has permitted the creation of important biological corridors and allowed the protection and increase of biological diversity. Through on-farm structured vegetation and canopy tree diversity, organic agriculture has demonstrated its capacity to provide livelihoods to farmers while providing easier movement of animals between managed forests and protected areas.

One of the largest, most unique conservation efforts presently underway in the world is the Meso-American Biological Corridor that runs across seven countries, covering an area of 770 000 km2, from Panama to Mexico. Examples of organic cultivations include: shade coffee in the buffer zone between El Imposible and Los Volcanes National Parks in El Salvador where hundreds of endemic and threatened birds, mammals, reptiles and trees strive; and shade cacao and bananas in the Talamanca-Caribbean Biological Corridor and its buffer zones, Costa Rica, where most of the fauna is endemic. Organic systems offer habitats similar to intact forests, providing suitable refuge for migratory and resident birds and foraging for many forest mammals, including species at risk.

5

Organic Agriculture Policies

In developed countries, farmers and consumers' demand for environmental and health quality created the organic agriculture movement. In USA, private and state schemes for certified organic food proliferated to a point that authorities were requested to establish federal rules and control systems to guarantee consumers' confidence in organic products.

In the European Union (EU) the demand for sustainable agriculture, and organic agriculture in particular, represented a perfect match to governments' priorities to reduce surplus food. In fact, EU organic agriculture policy reconciles agricultural and environ-mental policies as it represents a viable option for extensification and an alternative to land set-aside.

The present EU framework for organic agriculture aims at developing local economies and overcoming problems of discontinued supply through its support programmes. Although commitments to support organic agriculture vary between EU countries, the main objective is the move towards a model of sustainable agriculture and rural development. Active government policy not only supports organic farmers but (in some cases) requires it through established targets.

In developing countries, policies for organic agriculture seek to earn, through exports, foreign exchange for other development needs. In fact, declining government budgets have forced many developing nations to restructure their agriculture sector. Agricultural institutes and extension agencies have to provide more market-oriented services. Liberalisation and privatisation policies open the way for a greater role for entrepreneurs and producers' organisations. These trends are an impulse for private initiatives for organic agriculture.

When a critical mass of practitioners is formed for organic agriculture, governments formulate policies to support the marketing of certified organic products. The incentive of such policies is therefore economic, either for tapping lucrative markets, securing a place in world trade and/or counterbalancing withdrawal of government support to agricultural inputs and other services.

While the environmental and economic benefits are generally perceived in both developed and developing countries, there are reservations as regards the ability of organic agriculture to respond to other societal needs, namely to secure food needs. The review of the Commission on Sustainable Development (CSD) of the agricultural sector revealed further concerns of policy-makers with regard to organic agriculture.

The CSD-8 Report mentions the following: "while organic farming had been promoted and used in a number of countries, some countries were cautious on this issue. They felt that there could be limitations and risks in the organic production of food in developing countries, and that this approach should not be considered as a solution for developing country needs. The use of organic agriculture as a basis for setting standards that could lead to trade barriers should be avoided. Nonetheless there are a number of organic agricultural techniques that could be applied to enhance traditional and other agricultural practices to promote sustainable agriculture and rural development. The international community should assist developing countries to use and take advantage of these techniques".

Market Demand for Organic Agriculture Products

In both developed and developing countries, organic agriculture policies have developed bottom-up, emerging from societal pressure. National and supranational policies for organic agriculture have been so far concerned with creating favourable structures for certified organic products. The type of agricultural development (and supportive policies) is increasingly dependent on the market-economy. Market demand for organic agriculture products pulls agriculture production policies towards a more sustainable direction and links demand and supply.

Organic Agriculture Policies

Present and future policy development for organic agriculture will always have multiple objectives: income generation, natural resources conservation, food self-reliance and rural development—but with different levels of emphasis. For the sake of convenience, the examples presented below are grouped under one overriding objective (or entry point) but this does not exclude that a policy can target multiple objectives.

Organic Agriculture Generates Incomes

Organic agriculture generates incomes through international exports or by saving production costs. The objective of export promotion of organic products is not only driven by interesting premium prices but also by certain countries' search for a place on world markets. The objective of reducing costs through organic agriculture is achieved by promoting economic self-reliance and by saving on external input costs.

International markets

Issues: Institutions and agreements governing trade such as the Uruguay Round Agreements, regional trade agreements and preferential trade agreements are critical factors for competitiveness on the agricultural market. Trade liberalisation and the erosion of preferential trade agreements are stimulating the production, in developing countries, of value-added food products to secure markets for what is often their single export commodity. In particular, countries with small economies rely on a small range of commodities such as sugar cane, bananas and tropical beverages (coffee, cocoa, tea), many of which having suffered from a long run decline in real world market prices and slow growth in demand.

The global demand for agricultural raw materials also expanded less rapidly over the past two decades, due to the growth in the use of synthetic substitutes, but recently consumer demand of "natural" cotton textiles is increasing. Organic agriculture offers a specialised market and an opportunity to diversify into new commodities that have a high demand and interesting price premiums.

Countries with small economies: In small island developing states, erosion of trade preferences associated with sugar and bananas are of overwhelming importance, especially as the EU banana regime (which provides duty free access for ACP bananas) has been challenged by the World Trade Organisation. Banana producing countries are likely to be hurt if they fail to improve their competitive position and diversify into higher value products. Policy responses to emerging challenges have not materialised but the private sector in several small countries has already started converting into organic production.

Bananas: Markets of organic bananas have developed only recently and still represent very small volumes but imports have been growing at approximately 30 percent per year. The main supplying country is the Dominican Republic, followed by some other Latin American countries and the Philippines. Prospects for increased organic banana marketing are promising, especially as price premium at retail level may vary from 50-200 percent.

Although market demand is significant, organic bananas will remain a niche market as long as the considerable production constraints are not overcome (in particular Black Sigatoka and soil fertility aspects). In this respect, they may offer good opportunities to small and medium scale producers in countries where Black Sigatoka is absent or less prevalent. Even so, other production and shipping constraints will have to be solved.

Cotton: Several African and Latin American countries export certified organic cotton. The economic opportunities of this specialised market attracts the interest of private traders (US companies for naturally coloured cotton in Arequipa, Peru), national NGOs, international NGOs (assisting producers in Uganda and Senegal) or federal governments.

Policy requirements: In many tropical countries, limited supply besets organic agricultural growth. Producers and operators would benefit greatly from government policies that support research (to resolve production constraints) and conversion and certification costs, with a view to improving supply. Agricultural policies have not yet integrated organic agriculture policies as a means to meet the challenge brought by trade liberalisation.

Export promotion

Issues: The economic opportunities offered by organic markets in Western Europe, North America and Japan have attracted the attention of many developing countries. Favourable government intervention and support exist in a few developing countries (namely in Latin America) and steps are being taken by many others to support the export of certified organic commodities.

The apparently long-term potential of organic markets is not only fuelled by consumer demand and willingness to pay premiums but also by aggressive marketing and promotion undertaken by major retail groups. Most developing countries, however, obtain technical know-how and certification services from private organisations. The strong-market orientation of policy interventions for organic agriculture are evident in certain countries.

In India, for example, it is the Ministry of Commerce that registers farmers wishing to convert to organic operations but farmers are asked to seek (unavailable) technical assistance in the Ministry of Agriculture. In most cases, the development of organic agricultural exports of developing countries relies on private initiatives, which establish trade associations with foreign partners. This trend is also a creative response to the withdrawal of government support to agriculture.

Mexico: In Mexico the elimination of most input subsidies, rural credit and other Government support over the last decade pushed farmers to seek more profitable production alternatives. Counter-seasonal organic fruits and vegetables, herbs and medicinal plants and organic, shade and fair-trade coffee are exported to USA. Mexico is one of the leading exporters of organic coffee. Typically, small-scale farmers organise themselves into cooperatives, receiving support from rural development and environment NGOs. Joint ventures between US-based importers and small farmers organisations in Mexico are common, especially for the provision of specialised inputs, agronomist advice, and certification.

Turkey: In Turkey all certified organic production is destined to export markets. Ninety percent of products are dried fruits; the remaining includes nuts, textiles, medicinal plants and herbs. The main drive to this development was the Turkish Association of Organic Agriculture Movement (ETO), an NGO that today has an advisory role to the Ministry of Agriculture and Rural Affairs and related state offices. In 1994, the Turkish Government adopted the EU definition of organic agriculture and issued its own legislation as an opportunity to increase agricultural exports while providing income to marginal farmers.

Over the last few years the Ministry of Agriculture increased its policy interventions to support production and trade of certified organic products: a 1996 regulation requires Turkish exporters to declare certified organic goods exported to the EU; independent control and certification is authorised by the Ministry; extension staff is being trained by ETO on organic agriculture; a research working group on organic agriculture was established to set up priorities in research and to raise funds; and a special committee on organic agriculture was charged to coordinate and carry out activities on organic agriculture, especially with regard to collecting information related to certified organic products.

Tunisia: The proximity of a privileged organic market (EU) has triggered a relatively quick policy response from the Tunisian Government. Measures are being taken to encourage farmers' conversion to organic production while remaining competitive. In 1999, presidential measures were taken to comply with EU Regulation and a National

Commission for Organic Agriculture was established to encourage and stimulate the organic sector. A budget was allocated by the Ministry of Agriculture, including subsidies to cover 30 percent of investments of organic farmers and to cover 70 percent of certification costs over five years.

A certification authority (BIOCERT Tunisia) was created under the Institut National de la Normalisation et de la Propriété Industrielle. As there is a lack of organised sources for organic fertiliser supply, the Tunisian Institute of Appropriate Technologies is engaged in studies for composting organic waste of food industries and techniques to recycle olive water residues. A Technical Centre for Organic Agriculture is being created for professional training and to support research.

Guatemala: In Guatemala 35 years of civil war (where farmers abandoned their land) reclaimed the country's native plants such as cardamom and other spices and natural dyes. Market demand for traditional and chemical-free commodities represented a meaningful source of income to some peasants. Small-scale farmers and producers in northern remote jungle territories (namely Indian peasants in Las Conchas) became exporters to US markets under private projects that include organic cocoa, mango and equal exchange coffee.

Prospects: The so far insufficient supply of certified organic products can be boosted if technical production problems and certification costs are decreased: government policies play a key role in this respect. In particular, a legislative framework that provides definitions, standards and accreditation to certifiers is needed to protect responsible producers. Active government support to inspection and certification and market-oriented services are necessary to provide equal opportunities.

Effect of economic self-reliance

Issues: Developing countries are increasingly dependent on imports, namely cereals. Import substitution with local food varieties, can partly increase food self-sufficiency as well as dependence on volatile markets. Organic agriculture offers a means to substitute imported agricultural inputs with locally-produced organic inputs. In places where trade is constrained and foreign currency is scarce, organic agriculture becomes a solution to economic necessity.

Cuba: When the trade relationship between Cuba and the former USSR was severed in 1990, pesticide imports dropped by more than 60 percent and fertilisers by 77 percent. Pushed by economic necessity and in response to the crisis of agricultural inputs, Cuba adopted organic agriculture as part of its official agricultural policy. The Ministry of Agriculture and the Cuban Association of Organic Agriculture took far-reaching steps to promote organic agriculture systems and establish research programmes that laid the foundations for food self-sufficiency (namely fruits and vegetables) through organic management.

Among sophisticated experiments are bio-fertilisers, bio-pesticides and the use of fermentation and tissue culture. Investments are now being made to increase knowledge and technology for organic agriculture, through higher education curricula, with a view to create a new generation of agronomists.

Policy requirements: Cuba is perhaps the best example of large-scale government support to organic agriculture and is an encouraging model to replicate in other countries. It is an integral part of agricultural policy. Hence, organic agriculture development does not suffer from the disincentives that are typically brought by conventional agricultural policies. In fact, government-supported organic production does not have to compete with cheaper imported food, farmers are not subject to pressure for using synthetic agricultural inputs, and proper research and extension back-up efforts.

Synthetic pesticide reduction

Issues: Pesticide hazards to human health and the environment are controlled through increasingly stringent pesticide residue standards in food. Pesticide-use reduction policies are multiplying around the world and the decrease of subsidies on agro-chemicals represents an increase in production costs. Farmers are therefore reducing the usage of purchased inputs through decreased application frequency, transition to integrated pest management or conversion to organic agriculture.

Iran: In Iran farmers have so far obtained a 50 percent discount on cost of already subsidised pesticides. The existence of pesticide residues above internationally accepted standards for exported nuts triggered policy measures to tackle export difficulties. The newly established High Council on Pesticide Use Reduction Policies, chaired by the Ministry of Agriculture, decided to cut pesticide subsidies by 7 percent per year. This Council also commissioned a study on organic agriculture policies in the world to evaluate the feasibility of this system in the Iranian context.

Strict standards and the export orientation of organic agriculture were found to be less attractive than integrated pest management for the Iranian objective of pesticide reduction. The present implementation of non-chemical pest management in Iran is a reality that might potentially lead, at a later stage, to the adoption of broader organic agriculture policies. A National Committee on Organic

Agriculture has recently been established to consider the type of support that the Government could provide to peasants.

Egypt: The organic agriculture movement was born in Egypt some 20 years ago, chiefly to alleviate the increasing threat of pesticide poisoning to Egyptian farmers. Cotton cultivation is one of the most pesticide intensive crops. World-wide, 18 percent of chemical plant protection active ingredients are used in cotton fields which represent only 0.8 percent of cultivated areas. In the last two decades, the Egyptian average yield of raw cotton remained stable despite a continued increase of pesticides. In the early 1990s, SEKEM starting applying biodynamic methods to cotton.

The success in cotton pest control (by pheromones) raised authorities' interest in biological control: today, nearly 80 percent of Egypt's cotton cultivation applies biological pest control and the Ministry of Agriculture has forbidden arial sprays of pesticides on cotton, with a view to promoting biological control. In 1995, pesticide use in cotton dropped from 1 800 t to 320 t and average yield grew from 900 to 1 220 kg/acre. Organic cotton cultivation is based on intensive cooperation between farmers and scientists. The Centre for Organic Agriculture in Egypt operates an inspection and certification scheme according to the EU Regulation 2092/91.

Prospects: The various problems of synthetic pesticide sourcing and application create opportunities for organic agriculture development. Since integrated pest management has become a large reality in conventional agriculture, the target of zero synthetic pesticide use is being approached. The withdrawal of input subsidies also includes fertilisers. Regulatory measures and higher production costs are good reasons to increase fertiliser-use efficiency and to fertilise organically. As knowledge is gained in integrated production, the transition to wider organic agriculture policies will be facilitated.

Natural Resources Preservation

Issues: Environmental agreements and environmental concerns of society at large put a limit to agricultural expansion and mitigation of agriculture-induced land degradation. The Forests Principles, the Convention on Biological Diversity, and the Convention on Climate Change are examples of policy instruments that have a direct positive bearing on organic agriculture.

Agro-forestry and minimum tillage, that are common in organic systems, help carbon sequestration. Crop rotations and diversified production help biodiversity. Although organic agriculture for the sake of the environment is not the prime preoccupation in developing countries, some of these countries succeed in deriving livelihoods while protecting their natural capital.

Ghana: Ghana provides a good example of efforts to challenge the inherent low fertility of the natural environment through organic management. In the past decade, deforestation and inappropriate farming methods resulted in drying rivers and streams in the Techiman District of Ghana. Consequently, disputes over water were raised within the local Forikrom community.

A local young people's association was formed to help communities carry out local improvements and respond to natural disasters. With the help of the Department of Forestry, an afforestation project created a shelterbelt around water sources, a centre was created for forestry extension and nurseries were established for reforestation purposes and commercial farming.

The good sales prospects for cashew, palm, and teak, and the interest of the Environmental Protection Agency of Ghana, led to the creation of the Abrono Organic Farming Project to deal with the concerns of declining soil fertility that underfed the earlier conflict. Young farmers were trained in organic crop production and dry-season methods of vegetable gardening. Trainees have developed organic cashew farming and organised themselves into cooperatives to engage in block farming of 1-2 acre plots along the stream.

Today new sources of wealth have developed through new organic cash crops. The emergence of an environmentally conscious community serves as a model for other towns in the region. Shelterbelt with teak and leucenea have developed and water is abundant. People no longer burn the bush carelessly and there have been no bush fires since 1990.

In its efforts to halt declining soil fertility and to enhance productivity, the Ministry of Food and Agriculture and the Ghana Organic Agriculture Network is implementing a Sedentary Farming System project to enhance agricultural productivity. Organic farming is proving its viability as a better alternative for small-scale farmers who produce over 80 percent of locally grown food. The system is gradually reducing slash-and-burn to shift to intensified cultivation on a unit area. The Ministry of Food and Agriculture also supports organic rice-fish projects to enhance the ecological basis of local development.

NGOs and Government institutions are collaborating to cut pesticide costs, generate employment and minimize health and environmental hazards through organic agriculture. The Organic Farming Promotion Project is implemented by an NGO and a technical advisory board composed of officers from the Environmental Protection Agency and the Ministry of Food and Agriculture. Natural ways are promoted through research on compounded Neem extract concoction and garlic: trials on grain storage are carried out by the Plant Protection and Regulatory Services. Organic fungicides, foliar and fertilisers from local waste material are part of the research. Funding is however a major setback to further research and education.

Central America: In Central America most agriculture ministries are also responsible for natural resources. The Central American Commission for Development and the Environment provides a favourable political framework for natural resources conservation. Agriculture ministries possess the Action Plan for Central American Agriculture. Agricultural policies are presently directed towards export promotion of non-traditional products through intensive capital, labour and agro-chemical use. Existing non-tariff barriers to trade of many conventional agricultural products and institutional responsibility for both the environment and agriculture are favourable elements for eventual adoption of organic agriculture policies.

Prospects: It is evident that environmental conservation will remain second to livelihood needs in developing countries. Natural resources, however, are the basis for agricultural productivity. Organic agriculture has positive "side-effects" on the environment. Reconciling natural resource conservation and agricultural production through organic agriculture is, therefore, an avenue for supportive policies in developing countries. These policies should focus on facilitating research and education on agro-ecosystem management. Cooperation between environment and agricultural agencies on land use and standard setting is essential.

Poor Small Holders

Food self-sufficiency

Issues: Conventional agriculture tends to be unaffordable and unsuitable for resource-poor small holders. It requires inputs that are outside their reach. Subsistence farmers represent a large part of farmers in the developing world. For example, farmers marginalised by modern agriculture in India represent, according to some authors, about 70 percent of the farming community. Poverty and lack of production means undermine the productivity of traditional farming systems.

In developing countries, a good proportion of areas under cultivation produce below the "minimum acceptable level" due to different agro-ecological constraints and different agricultural management practices. Although the growth of global food production, in the next three decades, is expected to surpass population growth (1.5 percent and 0.7 percent respectively), 400 million persons will continue to suffer hunger in the year 2030.

Agricultural development depends on factors of production and on a sociopolitical environment dominated by poverty and food supply policies that create dependence on imports of cereals. As the problem of access to food, and means to produce it, are tightly linked to income growth, food security is likely to fall short of its goal as long as poverty persists.

The introduction of organic agriculture technologies in resource-poor areas has the potential of raising agricultural productivity while relying on local resources. Considering that the departure point is very low in many areas, especially in Sub-Saharan Africa, consistent efforts to enhance farmers' skills in managing resources and maximizing labour productivity can increase agricultural performance.

Such efforts do not mean only refraining from using external inputs but rather complementing traditional knowledge with results of modern agronomic and ecological sciences. The rehabilitation of soil fertility through organic management and diversified cultivation of traditional food can increase farming systems resilience, improve household nutrition, and decrease dependence on single crops for volatile markets.

Kenya: One of the main questions relating to the incorporation of organic agriculture into general agricultural policy is the lack of hard facts proving its actual performance in agronomic and economic terms. With the support of the Netherlands Ministry of Foreign Affairs, ETC-Netherlands and the Kenya Institute of Organic Farming conducted a four-year research to evaluate the future prospects of organic agriculture in low and medium potential areas in East Africa. Contrarily to the general belief that organic agriculture in the tropics is constrained by the insufficient organic material (to compensate for the use of external inputs), the study revealed a good performance of organic systems.

The organic system in medium potential areas significantly outperformed conventional methods in maize grain yields, net cash benefits, return to capital and return per family labour day. Organically grown maize experienced less weevil attack during storage than its conventional counterpart. The outcome of this research was shared with researchers, extension staff and farmers. Collaboration and further deepening of future research with mainstream institutions is being discussed.

In particular, there will be a need to explore site-specific techniques and to study the long-term effect of organic matter on soil fertility and nutrient balances. In Kenya, the possibilities offered by organic farming are particularly appealing to women. They offer the possibility of successfully securing food for the family, without having to depend on cash or potentially dangerous chemical products.

India: India has the second largest population in the world. Agriculture occupies 45 percent of the total geographical area and is the primary occupation of more than 60 percent of the population. A pre-colonial survey of 800 villages in South India indicated a prosperous average agricultural productivity. Colonialism destroyed the self-reliant nature of village economies and frequent famines led to the adoption of the Green Revolution technologies that met food needs and generated export surpluses. This laudable effort in reaching higher levels of production left, however, many side effects, which continue to impair the sustainability of Indian agriculture.

Use of chemical inputs, aggressive propagation of hybrid seed varieties, uncontrolled mining of water resources, and turning the agricultural sector to market economies disoriented the self-reliant economic system into a dependant economy that neglected staple food for poor farmers, crop rotations, tree growing, and other traditional practices. Fertile soils are today seriously degraded, water logging is a problem in irrigated areas, and depletion of soil nutrients is increasingly evident. Consequently, the productivity of most crops is affected and pesticide poisoning is recurrent among humans and animals.

A renewal and adoption of holistic traditional practices through organic agriculture is in its infant stage. Mahatma Gandhi pioneered organic agriculture through constructive programmes in several locations in India. A protagonist of self-reliance, he taught his fellow workers about composting and farming based on local inputs. The Gandhian concept of organic agricultural practices is spread over different parts of the country. In hilly regions, tribal areas and other marginal regions, many small farmers are de facto organic producers. Out of necessity they have turned degraded lands into productive organic systems that meet local needs. Surpluses are sold in village "green shops" at prices often equal to conventional products.

Many NGOs are today active in promoting organic agriculture in India. Growing environmental consciousness and fears of health hazards of conventional food has spawned domestic consumption of organic food. Expansion of domestic markets is leading to the surge of organic producers attracted by premium prices.

In the year 2000, the Horticulture Division of Department of Agriculture and Cooperation has created an organic farming cell to coordinate various activities of organic agriculture in India. For a few years now the Agricultural and Processed Food Export Development Authority has been promoting exports of organic agriculture products. The National Standing Committee on Organic Agriculture (set up by the Ministry of Commerce), collaborates with SEWAK to assess the feasibility of organic production and consumers' response on domestic markets.

Government interventions established so far by the Government of India focused on promotion of certified organic production, first for international markets and more recently, domestic markets. Organic farmers considered in surveys are richer producers who have market access. Resource-poor farmers practising organic farming have not yet been the focus of attention. Even for certified organic products, India still has a long way to go in developing the necessary policy framework and implementing it. There is a weak political will, propaganda for agro-chemicals erodes confidence in organic agriculture potential, small farmers are marginalised, there is a lack of a national body of farmers, lack of training, lack of national certification and lack of organised markets within the country.

Prospects: The interest in organic agriculture in developing countries is growing because it requires less financial input and places more reliance on the natural and human resources available. Studies undertook to date seem to indicate that organic agriculture offers a comparative advantage in areas with less rainfall and relatively low natural soil fertility levels. Labour realises a good return and this is very important where paid labour is almost non- existent.

Agricultural policies should revise their food supply strategies and valorise local production. Organic agriculture does not need costly investments in irrigation, energy and external inputs but rather substantial investments in capacity-building through research and training. Pro-poor organic agricultural policies have the potential to improve local food security, especially in marginal areas.

Wider Social Development

Issues: Urban and peri-urban organic production provides food to vulnerable groups. Developing countries' farmers easily adopt organic techniques because they are close to their traditions and culture. The potential of organic agriculture to generate employment and reverse migration to urban areas has not as yet been taken up by developing countries' agricultural policies. It seems, however, that this awareness is rising in economic and social government institutions in some countries.

Argentina: In Argentina a model of organic orchards is being experimented to promote employment and food self-sufficiency since the beginning of the 1990s. An NGO (pro-Huerta), with institutional support of the Secretariat for Social Development and other Government agencies, is seeking alternatives to improve the nutritional status of rural and urban populations and to create employment opportunities. This project benefited by 1996 nearly two million individuals, namely unemployed, indigenous, and food-insecure persons. The initiative promoted small-scale food production by providing capacity-building, technical assistance, development and validation of alternative technologies, inter-institutional coordination, and provision of critical inputs.

Costa Rica: In Costa Rica health motivations led to the formation of small-farmers groups to produce organic vegetables, which have been successfully marketed nationally over the last ten years. This movement was also stimulated by agronomic policies that withdrew credit preferences to small producers and decreased food prices due to foreign surpluses on the national market. Agro-chemical price increase, soil exhaustion, health hazards, and growing alternative markets motivated producers to convert to organic production.

Although market penetration is different according to products, organic production of broccoli, cassava, and hearts of palms is today competitive. Banana production includes ethical trading with the UK and ECO-OK bananas that are sold without premium. Recently, the Ministry of Agriculture and Cattle Raising has created a National Programme for Organic Agriculture. The Government objective is to develop a national strategy for organic agriculture to encourage the sector growth with a view to use natural resources sustainably and improve the quality of life of farmers and consumers.

6
Organic Agriculture For Adaptable Climate Change

Organic agriculture (OA) provides a broad set of practices that increase resilience in farms. OA is based on a rigorous standard requiring no synthetic inputs, instead using practices modelled on ecological processes to increase soil fertility and ward off pests. The Codex Alimentarius Commission defines OA thus:

> "Organic agriculture is a holistic production management system which promotes and enhances agroecosystems health including biodiversity, biological cycles, and soil biological activity. It emphasizes the use of management practices in preference to the use of off-farm inputs, taking into account that regional conditions require locally adapted systems. This is accomplished by using, where possible, cultural, biological, and mechanical methods, as opposed to using synthetic materials, to fulfil any specific function within the system"

This definition serves as the baseline for OA around the world. Countries develop their own standards from those set out by the Codex Alimentarius Commission. Although OA adheres to certifiable standards, farmers can do much more to base their practices in ecology than standards require. The principles of OA concern the way people interact with living landscapes, relate to one another and shape the legacy of future generations. The International Federation of Organic Farming Movements (IFOAM) describes four principles of organic farming, upon which its standards are based. Each of the principles contributes to the longterm health of the farm, the surrounding environment, and the farming participants, which in turn builds resilience for the long-term success of the farm. The principles are the following:

— "*Health:* OA should sustain and enhance the health of soil, plant, animal, human, and planet as one and indivisible.

— *Ecology:* OA should be based on living ecological systems and cycles, work with them, emulate them, and help sustain them.

— *Fairness:* OA should build on relationships that ensure fairness with regard to the common environment and life opportunities.

— *Care:* OA should be managed in a precautionary and responsible manner to protect the health and well-being of current and future generations and the environment".

Although OA is not designed with climate change adaptation as a primary goal, OA's systemic, ecological approach inherently includes positive side effects, including reduced greenhouse gas emissions and strengthened adaptation strategies, especially for small holders who stand to suffer the most from climate change.

Climate Change Adaptation and Mitigation

OA has a role to play in climate change adaptation and mitigation, including avoided damage, and many farming practices contribute to both processes. Agriculture is a major contributor to climate change globally, chiefly in terms of methane and nitrous oxide emissions from livestock production and soils. Since the 1990s, agriculture has been responsible for 15% of total greenhouse gas emissions worldwide, accounting for one quarter of carbon dioxide emissions, two-thirds of methane emissions and nearly all nitrous oxide emissions.

Should direct and indirect sources of carbon emissions be considered, such as processing and transportation of foods, it might well be that food system releases more greenhouse gases into the atmosphere than cars do! Agriculture also has the potential to avoid climate change through emission reductions and mitigate climate change through carbon sequestration.

While individual practices could be implemented on almost any farm, OA is unique in creating a whole system of agriculture based on ecological principles from production to consumption by privileging closed energy and nutrient cycles at the farm and by promoting short supply chains.

While agriculture's mitigating potential is not nearly enough to prevent climate change from happening, its potential to reduce climate change is quite significant, particularly in concert with reforms of industries aiming at reducing energy use and promoting renewable energy sources. OA is often ignored in discussions of climate change mitigation, but is worth considering, especially if the processes described below are scaled up through conversion of larger agricultural areas and more research.

In France, for instance, it has been estimated that a national conversion to OA could possibly decrease greenhouse gas emissions by 10% including: carbon sequestration (-4% from increasing carbon in soils of 7-10 million tonnes per year); nitrous oxide (-3% from organic soil management); methane (-1% from enhanced manure management); carbon dioxide.

Carbon sequestration:

- Organic soil management focuses on increasing soil organic matter, which increases carbon sequestered in the soil. Several studies have shown that feeding soils annually with organic matter in excess of the mineralised quantity is the only way to increase soil organic carbon. Organic practices that do so include addition of manure and plant residue to the fields, mixed cropping, green manuring, legume based crop rotations, agroforestry and minimum tillage.
- OA conversion in orchards often entails changes from annual cropping or bare soils on the orchard floor to more permanent plant cover that sequester more carbon.
- Agroforestry, often used in tropical organic systems, sequesters significant amounts of carbon in the trees. Depending on the region of the world and the harvesting rate, agroforestry systems can sequester between 9 and 63 tons of carbon per hectare.

Carbon dioxide emissions:

- OA encourages minimal clearing and burning of vegetation and developing more permanent cropping systems that return biomass to the soil.
- OA does not use energy-intensive synthetic fertilisers and pesticides, thus avoiding the fossil fuels used to make these products and to transport them to the farm.
- In organic livestock production, animal feed tends to be produced on the farm, saving fossil fuel use for shipping feed. IFOAM standards require that at least

50% of feed be produced on the farm or on organic farms within the region.

— In general, OA privileges local produce, minimally packaged and processed commodities and short supply chains, thus contributing to reduced energy consumption.

Nitrous oxide emissions:

— Nitrous oxide emissions occur from all uses of fertiliser. Total nitrogen concentration in the application and the plants' capacity to absorb it determine the amount of emissions. OA is less likely to over-apply nitrogen due to its use of less concentrated forms of fertiliser. The distribution of organic fertiliser throughout the year as well as the nature of the soil and its humidity have a role to play in the degree of nitrous oxide emissions. Still, the fact that nitrous oxide emissions are proportional to nitrogen inputs results in decreased emissions in organic systems as compared to conventional ones.

— Improved ruminant diets reduce nitrous oxide emissions. In one study, diets with lower protein and higher fibre contents result in reducing emissions by 10-15% . There may be trade-offs, though, between the production of nitrous oxide with respect to the meat and milk productivity of the animals.

— Livestock concentrations are kept low, so the manure produced is not too much to be applied to the land.

Methane emissions:

— Not all organic practices reduce methane emissions. Specifically, increased biological activity in the soil can increase methane emissions, actually putting OA at a disadvantage in this regard, especially in humid soils. Furthermore, organic ruminants may produce more methane due to extensive pastures. This is partially compensated for by the fact that organic milking cows live longer, thus producing less methane in their life span.

— Compost, which is recommended in OA, emits less methane than manure application because it ferments aerobically rather than anaerobically.

— Improved ruminant diets can substantially reduce methane emitted in the animals' flatulence, due to bacterial fermentation of feed in the animal's main stomach.

— More research must be done to reduce methane emissions from rice paddies in organic systems, but there are several possible means to do so.

Organic Agriculture and Climate Change

Great convergence between the Codex Alimentarius Commission and IFOAM standards, the latter International Basic Standards are used because these are revised by the organic farming community every three years and are thus more up-to-date. These standards represent a baseline from which farmers can build their own systems, often going much farther towards ecological principles.

Soil and Water Resources

Without healthy soils and sufficient water, no agriculture is possible. Changes in temperature, precipitation and other atmospheric conditions due to global warming will directly change soils. While specific impacts are incredibly difficult to predict for individual areas, expected extreme weather events will almost certainly damage soils and accelerate erosion. Higher temperatures may increase frequency of droughts and reduce areas under permafrost.

Extremes in rainfall may leave soils too wet or dry for tillage at crucial periods. Efficient irrigation systems will become a necessary component of farms in water-stressed regions.

Soil management

There are many opportunities to improve the quality of agricultural soils, leading to higher productivity and better responses to extreme conditions. Conventional agriculture relies on external inputs, namely chemical fertilisers, for soil fertility. OA relies on management of natural cycles to add crucial nutrients, increase soil organic matter and protect the soil from erosion. As crops deplete nutrients, the addition of nutrients restores fertility for future crops.

Soil organic matter determines much about the soil's quality as it "is an important substrate of cationic exchange, is the warehouse of most of the nitrogen, phosphorus, and sulphur potentially available to plants, is the main energy source for microorganisms and is a key determinant of soil structure". Soil organic matter also helps soils retain moisture. Rain and wind both erode soils, costing the farmer loss of fertility and contaminating local waterways. OA practices ultimately increase long-term soil productivity, lending greater stability to the system.

Good soil fertility and tilth form the basis of sound organic management, and farmers practice a variety of techniques to ensure soil quality. OA farmers develop complex crop rotations. These rotations incorporate crops with different nutrient needs, allowing maximum use of soil nutrients. Rotations additionally break weed and insect pest life cycles, reducing the interventions necessary for pest management. Rotations typically include a fallow period in which the soil rests; plant residues from what grows on the field during that time are tilled back into the soil to add nutrients and soil organic matter.

Farmers may specifically sow a leguminous cover crop during the fallow period, or incorporate a legume at some point in the rotation, which produces nitrogen to be used by later crops. Fields may be used alternately for crops and for livestock pasture; livestock consume weeds and naturally fertilise the soil with their waste. Materials that might otherwise be viewed as waste are crucial for nutrient management in organic systems. Once returned to the soil, crop residues and animal waste add nitrogen and soil organic matter. Waste materials can also be composted. Plant residues are often added to livestock waste and allowed to mature into manure, which better balances the nutrient composition before being returned to the fields.

Cover crops help hold the soil in place during fallow periods. Permanent vegetation, such as trees and hedgerows, on the borders of fields protects fields from strong winds and storms. Such vegetation also provides benefits for biodiversity. Shelterbelts in Africa have also been used to reclaim desertified areas by blocking hot winds and retaining moisture, and traditional seed holes called "zai" are used in the Sahel to restore arid and crusted areas of fields.

Water management

By increasing soil organic matter, soil management techniques become water management techniques as well. A study comparing fields managed under conventional and organic systems found that the organically-managed farmers fared significantly better during drought. Researchers speculated that organically managed soils had better water holding capacity due to increased soil organic matter. Interestingly, the same fields fared better during extreme rainfall, absorbing more water and experiencing less runoff and erosion. Water absorption is necessary for groundwater recharge, an environmental service that protects the whole ecosystem.

In the face of climate change, farmers may need to take extra steps to ensure water availability. Various techniques for water harvesting, infiltration pits to collect water, and the cultivation of drought-tolerant crops have all been used in Zimbabwe as a means to improve access to water. Permaculture, the design of holistic, edible landscapes adapted to the local area, is also common in many places. Mulching helps retain soil moisture and ultimately adds organic matter back to the soil. Ultimately, irrigation may be necessary to adapt to reduced rain in certain areas. OA practices form the basis for protecting as much available moisture as possible in the face of challenges in finding sufficient quality and quantity of water for irrigation.

Biodiversity and Landscapes

As temperatures increase and weather patterns shift, ecosystems will change drastically. As the climate changes, habitats will become less suitable for some species and more suitable for others. These changes will happen across biological kingdoms and likely encourage migration of species to more suitable locations for their survival. In some cases, habitats will become like other known habitats, but others may be completely new. In agriculture, these changes may mean certain crops can no longer be grown where they currently are while others will be more appropriate. Weeds, fungal diseases, beneficial and pest insects all may change range, creating new situations for farmers to address.

Agrobiodiversity

In order to determine how an agroecosystem will respond to changes, one must ask to what extent the agroecosystem mimics Nature. Fields of genetically-identical crops facing a previously unknown pest may suffer great losses; history has recorded such devastations as the Irish potato famine and Southern corn blight in the U.S., both caused in part by the genetic uniformity of the crops. Biologically diverse systems are much more complex and rely on many relationships within the system. As climate change occurs, the ecosystem on a diversified organic farm is more likely to go through natural stages of succession, adapting in ways that prevent whole agroecosystem collapse.

Conversion of land to agricultural use has modified habitats for species around the world, and increasing population pressures cause agriculture to compete with wilderness as a land-use in many ecologically sensitive regions. Simplifying farms by growing monocultures and removing vegetation in the margins further reduces biodiversity in and around farms. Such practices not only cause environmental damage but create agroecosystems that will be less resilient to climate change. Organic farms, on the other hand, are designed with biodiversity in mind, as diverse ecosystems provide a number of important services on the farm. It is important to note that these services are developed on the farm and rely on farmers' knowledge, not external inputs, making such approaches particularly valuable in developing countries.

Agricultural biodiversity includes not only the genetic resources of crops and livestock, but the diversity of species that support production and those in the wider environment as well, such as soil microorganisms, pollinators and pest predators. When farmers design practices to enhance biodiversity not just on the farm but in the surrounding ecosystems, it can strengthen both the farm and the environment.

OA utilises various combinations of plant and animal species in time and space. Some organic systems are more diverse than others. Crop rotations are a requirement of organic standards. Polycultures are not required on organic farms, but certainly fall within organic standards and are quite common on farms in developing nations, especially in tropical climates. Polyculture on small-scale farms and home gardens is widespread throughout Central America, Southeast Asia, and sub-Saharan Africa. Tropical agriculture in particular includes complex agroforest polycultures with multiple year rotations and long fallow periods. Organic shadegrown coffee is an example of an effective means of using OA for agroforestry.

Organic standards also require farmers to use organic seeds and encourage genetic diversity in both crops and livestock. Organic farmers, particularly those in developing countries who cannot afford to buy their own seed, rely on saving their seeds year after year. Farmers select seeds from successful plants and in so doing develop local landraces, defined as "geographically or ecologically distinctive population which are conspicuously diverse in their genetic composition". The combination of these efforts increases farms' resemblance to natural ecosystems, thereby enhancing resilience.

Crop rotations are the most basic and common form of diversity on organic farm. As discussed above, organic farmers rotate crops and livestock from one part of the farm to another rather than grow the same product over and over in the same space. This practice provides multiple benefits for soil quality, and also helps to break pest and disease life cycles and maximizes efficient use of soil nutrients and water. Some farmers also use multiple sowing dates, thereby lessening the chance that the whole crop will be at a critical stage during an extreme weather event.

Livestock can play an important role in nutrient cycling. In Peru, cattle graze on crop residue, meanwhile fertilising the land with their waste, as part of a long rotation that includes quinoa, maize, alfalfa, potatoes, peas, wheat, and other crops. Most crop rotations include a fallow period in which a piece of land is allowed to rest. In the humid tropics, such fallow periods are particularly important as they break pest life cycles that winter would disrupt in temperate regions.

Tropical agroforestry includes rotations that may last several years. Fallow periods are often longer in these systems, allowing succession to occur as in a natural forest and contributing vital nutrients to soils often subject to leaching. Practices that increase soil health and reduce pests make crops healthier and thus more able to withstand changes in climate.

Diversity in space involves polyculture or intercropping. Simply put, these practices involve growing more than one crop in the same field. Such fields may involve alternating rows of two or more crops, or a crop grown underneath trees. Polycultures may also be so complex as to be indistinguishable from the natural landscape to the untrained eye. Polycultures reduce risk, for if one crop fails, there are still several others which may succeed.

In the case of an extreme weather event, some crops may be more able to survive than others, and different crops may respond better to drought versus a storm, for instance, and vice versa, helping to manage variable conditions. Interestingly, well-chosen polycultures tend to increase productivity. Not every crop has the same nutrient and moisture needs at the same time, so the plants use resources more efficiently than if only one crop is planted in the same area. Thus, the total productivity of all the crops grown in one area is often higher than if only one crop had been planted.

A diverse landscape offers many more opportunities to manage pests. Polycultures can distract pests, as the field does not offer the feast that a monoculture does, and organic farmers often place certain plants around the borders of fields that repel pests. Pesticides used in conventional agriculture often cause collateral damage which can then not be relied upon to enhance the agroecosystem. Organic farmers, on the other hand, design their plantings to attract pollinators and pest predators. Farmers may include plants that are not useful as edible crops simply because of their important role in pest management.

These plantings combined with natural vegetation on the edges of fields and woodlots create a habitat that can support beneficial species and the wildlife of the area in general. Because of the system's diversity, polycultures are less vulnerable to new pests or to the loss of a beneficial insect, as is likely to happen as habitats shift under the influence of climate change. Overall, diversity in the agroecosystem supports diversity in the surrounding environment, which increases the larger ecosystem's ability to cope with climate change as well. Given the importance of environmental services off the farm, practices that encourage resilience in the whole landscape are important indeed.

Not only are whole agroecosystems diverse, but the very crops and livestock grown represent many generations' worth of selecting seeds and breeding livestock that survive well in a given area. Thus, farmers have developed unique landraces specifically adapted to survive under local conditions. The diversity of local landraces is much higher than modern cultivars. Such diversity and adaptation certainly increase resilience on the farm. While modern varieties engineered for one or maximum two desirable genetic traits may offer the opportunity of high yields with heavy use of external inputs, such varieties are more vulnerable to changing conditions. Local landraces, on the other hand, tend to offer the stability of more consistent yields even in bad years. The following examples illustrate how farmers have revitalised the use of landraces to increase resilience:

— A group of Indonesian farmers, known collectively as "Pusspaindo" became frustrated by the environmental and health effects of conventional rice farming and have returned to using their own varieties under organic management. The farmers have found this change to be easier and more economical, as well as offering a wider array of tastes in rice..

— Peruvian farmers are growing genetically diverse cotton varieties that naturally produce a range of colours. The genetic diversity helps improve their farming operation, and the environmentally-friendly production methods appeal to consumers, helping ensure a market.

— South African farmers, faced with an outbreak of Newcastle disease in their chickens, found that their original breeds were more disease-resistant and also foraged a wider variety of food and survived better under extreme conditions.

Seed banks around the world collect varieties of seeds for the use of future generations. OA agriculture helps preserve not only the seeds, but the practices that required to grow them. As climate change occurs, it will be of most benefit to have the widest array of genetic diversity in crops and livestock possible. Thousands of varieties and species across the spectrum of edible plants and livestock have been lost in the last century, and more must be done to preserve these varieties and encourage their use and further development.

Agricultural biodiversity in time and space increases resilience in a myriad of ways: complementary use of soil nutrients and water, increased total productivity through appropriate polyculture mixtures, decreased risk from one crop failure, pest protection, the creation of microclimates suitable for beneficial insects and strengthening the genetic traits of local landraces. Evidence is beginning to show that "local practices of mixing species and varieties have beneficial effects on crop production over time, especially by buffering climate variability and reducing pest damage".

Landscape management

Organic farms often include diverse landscapes, which require special forms of management, and management techniques in the community as a whole can also benefit the entire ecosystem. No organic farm operates separately from other farms around it and the rest of the ecosystem. Every farm depends on interactions within the ecosystem, and briefly consider the role OA plays in the resilience of the surrounding environment. Organic standards have requirements in this area, including a prohibition on clearing primary ecosystems and several recommendations to include wildlife refuges and habitat corridors on farms.

Some polycultural systems go so far in mimicking natural ecosystems that "agriculture becomes a sequential harvesting system of crops and non-food crops". Traditional agroforestry often leaves the forest largely intact with gaps to produce crops or promote the growth of wild crops. Understory cropping systems also modify the existing habitat rather than replace it altogether. Shakedown coffee, for instance, is often grown in this way, though it can also be grown under trees planted for the purpose. All of these systems make organic farms connect more naturally with wilder land around it and cause less habitat disruption than conventional farming would.

For farms with less complex polyculture systems, it is particularly important to examine the fields as a whole. While land unused for production may appear unimportant, these places offer several opportunities to increase biodiversity. Hedgerows and vegetative buffer strips, essentially permanent vegetation such as shrubs and trees on the edges of farms, provide habitat for a variety of creatures and offer corridors between one section of habitat and another. Riparian corridors are also important; planting native vegetation on the edges of streams and irrigation ditches can offer habitat and prevent soil erosion as well. Riparian corridors, ponds, and even wetlands such as rice paddies also provide important habitats for migratory fowl and watering holes for other animals.

Herders must consider their impact on a large area of landscape. Sahelian herders use smallscale movements based on ecosystem feedback. African herders are known to maintain range reserves, areas they do not grase that provide resources in drought years. Such buffers help improve the resilience of the system to disturbance.

Community Knowledge Systems

Farmers worldwide hold a huge body of information about their farming systems. Farmers have observed phenomena and adapted their farming systems to better suit changing needs. Such practices are particularly seen in the array of crop varieties and livestock breeds developed solely for local use. The risk of global climate change is that changes will accelerate such that farmers will need to observe, learn, and respond more quickly than before. Changes will be faster and more radical, perhaps requiring new means of sharing information. OA is based on ecological processes; knowledge of the agroecosystem is thus a prerequisite to any organic farm. Farmers with a traditional knowledge base stand at an advantage in further developing ecological processes to respond to the effects of climate change.

Seed varieties developed to assist developing nations often do not succeed in the absence of large amounts of inputs and are less adapted to stress conditions. Locally adapted varieties are better able to handle difficult climate conditions. Traditional knowledge extends beyond crop and livestock varieties to apply to every aspect of the farming system. Most traditional practices seek to avoid risk and attempt to maximize the use of local resources. Monitoring the status of and changes in ecosystems is essential to this process, and keen observations will become yet more important as climate changes. Many communities use wild species as indicators of agroecosystem health.

Specific practices that confer resilience and health to the ecosystem may not be expressed as such, but rather as cultural practices and taboos. Some communities, for instance, protect important beneficial species from being hunted, collected, or disturbed during vulnerable life stages. Communities rely on local institutions and leaders to enforce these rules and also to coordinate appropriate resource use. Sustainability can thus be embedded in the very framework of local culture.

As knowledge tends to be based in communities, not just individuals, farming communities that share information build resilience by finding and disseminating best practices and coping mechanisms more quickly. Farmers must transmit information to each other and to the next generation for these systems to survive. Some of the best examples of knowledge transmission actually come from communities that stopped using many of their traditional practices and are now in the process of reviving them. In the central highlands of Mexico, farmers facing severe erosion have created the non-profit group, Vicente Guerrero Group (VGG) to revitalise sustainable practices to improve their soils. VGG trains farmers to teach other farmers best practices, emphasizing that farmers have an obligation to share their knowledge with each other.

This group has trained over 2 000 farmers in soil and water conservation practices in Mexico and Latin America. The Indonesian group "Pusspaindo" has instituted extension workers in every participating village. The extension workers are responsible for maintaining audits of natural resources in each village and networking with other extension workers and farmers about problems, such as the potential loss of a local landrace.

The development of cooperatives for organic marketing has helped facilitate the sharing of organic practices. A different group of Indonesian farmers collaborated with local NGOs and the National Parks of Indonesia to create the Indonesian Cassia Cinnamon Project, a cooperative to gain better access to markets for their organic spices. The cooperative then reinvests in community improvements, and the trees used for spices serve as a buffer habitat for a national park. Nearly 3 000 farmers have joined the cooperative.

As farmers develop more accountability towards each other and their own standards within organisations or marketing groups, they also create support networks to improve and extend organic practices. Community cohesion has proven to be a necessity to decrease certification and marketing costs as a critical mass is developed at all levels, from the cumulative knowledge generated to the quantities of produce grown. Having existing communication systems, coping mechanisms and commitment to ecological principles should also help farmers respond cooperatively and collectively to global climate change, increasing their adaptive capacity as a community.

Community knowledge as a form of adaptive management

Traditional knowledge is not just a system for the present, but a source of institutional memory about what practices have worked best over time. Such knowledge has been described as a "reservoir of adaptations," a whole set of practices that may be used again if the need arises. Likewise, farmers do not just observe, but create their own experiments and trials to improve their techniques. Not every traditional practice is ecologically sound. Some practices may have been useful at one time, but are not appropriate due to new conditions or overuse. Community knowledge thus represents a process of learning as much as a single body of information.

This process of building traditional knowledge has been compared to the scientific approach of adaptive management, so much so that adaptive management has even been described as science's rediscovery of traditional knowledge processes. Adaptive management is different from other modes of scientific inquiry first and foremost because it is based in solving ecological problems, which are often too complex for one to have all the information at the outset but yet also too urgent not to respond.

Adaptive management requires willingness to try an approach, evaluate it keenly, and modify the approach in an iterative process; in other words, it accepts the inherent uncertainty involved in manipulating whole ecosystems. Rather than attempt to change an aspect of an ecosystem, adaptive management focuses on improving the resilience of the ecosystem as a whole, which is a good basis for using adaptive management to help farmers respond to climate change. Traditional knowledge systems and adaptive management are both by necessity interdisciplinary and include a variety of players. These systems involve learning "not at the level of the individual but social learning at the level of societies and institutions".

There is much to be learned from indigenous organic agricultural practices that can be applicable elsewhere as farmers worldwide respond to global climate change. As scientists study these systems, and increasingly work with farmers to improve them, scientists do so on the farmers' own terms. Using adaptive management best resembles farmers' existing means of using observation and trial and error to improve their farming systems. One interesting example of a participatory adaptive management approach is the reintroduction of local pumpkin landraces in Cuba.

In response to a shortage of inputs, the Cuban government began several research projects to find alternatives means of production. After trials with conventional seeds, pumpkin yields decreased significantly. Researchers decided to try local pumpkin landraces instead, involving farmers in growing a wide array of varieties and selecting those that gave the best yields with low inputs. Farmers were able to select genotypes that worked best in their real-world conditions, and these seeds were used to supply other farmers. The project was so successful that Cuba has developed a programme called Participatory Plant Breeding for Strengthening Biodiversity with several other crops.

7

Handling and Processing of Organic Fruits and Vegetables

The value of the world organic market was estimated to be over US$ 15 billion in 2000. The USA and Germany dominate this figure. Despite this apparent large scale, the estimated share of organic products in total food sales is around 1%; the highest estimated share is in Denmark, where organic produce is estimated to reach 3% of total food sales. Organic fruit and vegetable sales account for around 20% of the total organic sales and represent around 2-10% of all fruit and vegetable sales in the developed countries for which estimates have been made.

The organic market follows the trends of conventional products. Bananas and citrus fruit are generally the largest volume fruit imports. Other principal fruit imports are pineapples, mangoes, and avocados. The vegetable market is generally less developed.

Unfortunately for developing countries, affluent consumers in Europe, USA and Japan prefer organic products from their own country or region. This partly reflects a distrust of other country's certification systems; but in part may be a reaction to the view that transporting foodstuffs internationally is wasteful of fossil fuels. The

UK and Belgium have smaller domestic organic production and consumers show less difference in trust between domestically-grown and imported organic products.

Non-organic Production

Many post-production operations for organic produce are identical to non-organic production. Where there are particular restrictions or considerations they are identified. The draft IFOAM Basic Standards 2002 state: 'Handlers and processors should handle and process organic products separately in both time and place from non-organic products. Handlers and processors should identify and avoid pollution and potential contamination sources'.

Harvesting Organic Procedure

There are few specific considerations for harvesting organic produce. Normal attention must be paid to harvesting each crop at its optimum maturity, bearing in mind its intended market; harvesting early in the day and keeping harvested produce in the shade, wherever possible; and removing field heat quickly.

Curing Organic Procedure

Some root, tuber and bulb crops require a 'curing' period at ambient or elevated temperature to promote wound healing and ensure optimum storage life. There are no specific requirements for curing organic produce.

Packing House Operations

Most markets require strict attention to be paid to the size, grade, quality and maturity of produce, whether or not it is organic. Fruit and vegetables must be cleaned and graded to comply with these regulations. Special consideration needs to be paid to the cleaning or sterilisation of grading and processing equipment in an organic operation: organic produce must be 'free of substances used to clean, disinfect, and sanitise food processing facilities', which can be achieved by re-washing equipment with hot water after the use of cleaning agents, or passing non-organic produce through the system before the organic produce. Ethylene is permitted for post-harvest ripening.

Packaging Materials

The IFOAM Draft Basic Standards 2002 state that 'Organic product packaging should have minimal adverse environmental impacts'; and recommend that 'Processors of organic food should avoid unnecessary packaging materials; and organic food should be packaged in reusable, recycled, recyclable and biodegradable packaging whenever possible'.

Thus, although all types of packaging are authorised, there is an expectation that careful thought will have gone into the choice of the packaging with regard to its environmental impact. In the future, restrictions may be put in place concerning the use of packaging materials that are harmful to the environment, especially for those packaging materials that are not recyclable or biodegradable.

Cardboard and paper

Whilst these traditional materials are generally readily available and inexpensive, they have several drawbacks: porous to gas, permeable to water, easily torn or crushed. They protect products only from light impacts. In organic farming, these materials are principally used for fresh fruits and vegetables. In order to limit impacts between products and to limit movement within the packaging, the use of liners between layers of fruits and vegetables, or of individual paper wrapping, can be efficient. Waxing of packaging restricts water permeability but can make the package unsuitable for recycling. Particular consideration needs to be paid to the use for which the packaging is intended.

A lightweight cardboard box may be adequate for use in the local market, but a telescopic box with reinforced corners may be necessary for seafreight. It will need to retain its strength during extended periods at low temperatures and high humidities to enable stacking over 2 m high on a pallet.

Plastic packaging

Plastic packaging is likely to deliver the best quality produce, minimising wastage. It can be pre-printed for marketing purposes and it is ideally suited to forming flexible, unbreakable packaging matched to the product's needs. It is light in weight, leading to cheaper transport costs and less fuel consumption in transport. Although plastic packaging is often frowned upon because it is commonly derived from fossil fuels and is not always able to be recycled, the alternatives need to be considered carefully. There are many recyclable or bio-degradable plastics and some plastics are now produced from starch.

Plastic is selectively permeable to gas and water, depending on the type of polymer. Some polymers are therefore ideally suited to creating a modified atmosphere around fresh fruits and vegetables. During storage, the respiration of the produce emits carbon dioxide and consumes oxygen. The selective permeability of the polymer results in an increase of carbon dioxide inside the packaging. The respiration rate of the vegetable decreases proportionally to this enrichment, and the composition of the atmosphere stabilises progressively. The resultant atmosphere, if maintained, can lead to significant improvements in storage quality.

This type of modified atmosphere packaging (MAP) is being developed for highly perishable, high-value produce. The problem is that temperature fluctuations dramatically affect the rates of tissue metabolism and the permeability of the plastics; stable gas compositions can be obtained only in a precisely controlled environment.

The risk of anaerobism in MAP is severe; anaerobism leads to the production of off-flavours and, in extreme cases, favours the growth of toxic organisms and toxin production. MAP is thus too risky to recommend for routine use in the export of fresh fruit and vegetables from developing countries.

Glass receptacles

Glass receptacles are principally used for liquid products or solids in liquid. Glass receptacles are well adapted to organic products as they are impermeable to gas, air moisture, microorganisms and resistant to thermal treatments. Against these positive attributes are the bulk and transparency of glass, which can cause problems for products that are sensitive to light. Also, the energy costs of recycling glass are very high (possibly making it less ecologically-friendly than plastic).

Metal

Metal offers the same sealing advantages and resistance to thermal treatments as glass. However, the consumer image of this type of packaging is unfavourable; and metal can be subject to corrosion.

Effect of Cooling System

It is highly advised to place harvested fruits and vegetables as soon as possible in a storage area that is kept at the appropriate temperature and to retain that temperature thereafter. Research has enabled the optimum storage temperatures to be established for many products. These guidelines need to be verified for each product, given the degree of harvest maturity, the variety, and the growing region of the produce. There is a range of cooling techniques available.

Air-cooling system

Air-cooling requires the supply of cold fresh air (RH 85-90%) to bulk or packaged products. This system requires a moderate investment. Controlling parameters is easy and the system is modular; however, the cooling process is very slow and not homogenous. For regions without access to electricity, very low-cost cool stores can be built which rely on evaporative cooling; water evaporation from e.g. wet sand between porous brick walls can lower the temperature inside a store to 10-15°C below ambient. Storage in pits or caves is a traditional means of taking advantage of the cool and more constant temperatures below ground. Night air can be used to lower the temperature of a well-insulated cool store, which is then sealed for the daytime.

Forced-air-cooling system

The simplest design is achieved by building parallel stacks of palletised cartons in a refrigerated cold room. The gap between the two parallel rows of pallets is closed off with a cover. A small fan is placed at one end. The exhaust fan removes air from the enclosed space, so that the pressure falls. Cold air then flows through the ventilation slots in each carton. Advantages of forced-air cooling are its capital cost, its flexibility (cartons, bins, before or after packing) and lack of condensation. Forced-air cooling is more rapid and even than air-cooling but not as rapid as hydro-cooling or vacuum-cooling.

Hydro-cooling system

Hydro-cooling involves immersing the produce in cold water. The advantages of this method are speed, uniform cooling and no weight loss by dehydration. Disadvantages include the necessity of drying the product surface after cooling and avoiding a build-up or transmission of disease in the hydro-cooling water.

There are also problems with the requirement for a large quantity of clean water, disposing of waste water, heavy capital cost and it is not applicable to all types of packaging especially cartons. This technique is used for small fruits or vegetables, leafy vegetables and pineapple. Another system which uses less water involves passing the produce through a cold mist at about 5° C.

Vacuum-cooling system

Vacuum-cooling is particularly effective for leafy greens The products are placed in a chilled vacuum chamber and air is exhausted by a vacuum pump. When the pressure is low enough the water in the produce starts to evaporate and cools the tissue. Pre-cooled fruits and vegetables are then placed immediately into a cool room. The decrease in temperature is very rapid but this technique is capital-intensive.

Fruits and Vegetables Storage

Organic product needs to be stored and transported in a way that it is properly identified and physically separated from non-organic products. Although individual products have a range of optimal storage temperatures, in practice most produce can be stored at one of three temperatures.

These recommendations should always be verified under local conditions for each variety and harvest maturity. For certain fruits and vegetables cooling in stages plus intermittent warming allows the produce to resist chilling injury and spoilage.

The refrigeration of fruits and vegetables improves when carried out with the proper relative humidity for each product. Appropriate storage conditions are always a compromise; high humidity levels limit dehydration and loss of weight, but encourage the development of microorganisms leading to rots.

When different kinds of produce are mixed in storage, consideration must be given to aroma volatiles which may taint other produce and ethylene which may ripen or damage other produce. The risk of mutual contamination is most pronounced between fruits which emit large quantities of ethylene and produce which is sensitive to ethylene like avocados, bananas and papayas.

Transporting Organic Fruit and Vegetables

There are no particular legislative requirements for transporting organic fruit and vegetables, air travel is extremely wasteful of fossil fuels. This is less of an issue if the produce is travelling with passengers but is a serious argument against the use of cargo-only planes for fresh fruit and vegetables if there are locally-grown equivalents available in the importing country.

Sea transport is many times more economical of fossil fuels (per km) than air- or even road-freight. It is worth noting therefore that produce grown under low-input agriculture in a developing country and transported by sea freight can have a lower 'energy overhead' than produce grown in Europe under intensive mechanised agriculture with high fertiliser inputs.

It is important to look at the infrastructure for freight before establishing a perishable crop as every delay in transit to urban centres, ports and airports reduces the potential shelf life of the product. Ideally refrigerated trucks should be used for produce stored at low temperatures to maintain the 'cool chain' from the grower's property to the marketplace. Harvesting should be timed to coincide with air- or sea-freight opportunities if refrigerated storage is limited.

Processing Method

Freezing process

Freezing is the only processing method that keeps produce in a state similar to the fresh crop. It is quite often applied to vegetables but rarely used for fruits, as they do not handle it well. If properly frozen, stored and defrosted, frozen produce becomes a high quality raw material. Nutritional quality is generally retained at at-harvest levels when the product is sold frozen. Significant vitamin loss may occur during subsequent thawing and cooking. Colour, odour and taste are retained well by freezing, but superficial dehydration occurs.

Before freezing, the fresh produce is cleaned and graded before being conveyed to the peeler, stoner or slicer. At this stage fruits and some vegetables such as onions and peppers are ready for freezing, but most vegetables need to be blanched with hot water or steam at 80° to 100° C to inactivate enzymes that could otherwise lead to a loss in vitamin C and flavour.

Fruit can be coated in sugar or in syrup that contains an antioxidant like ascorbic acid. Coating retards browning, avoids the "cooked" taste after defrosting and increases product quality. The products may be packaged before or after freezing. The procedure requires very rapid freezing. The degree of freezing depends on the duration of storage. Some conditions of freezing are listed in Table 1.

Table 1: Practical storage life of frozen products.

Products	Practical storage life (month)		
	-18°C	-25°C	-30°C
Fruits in sugar	12	18	24
Fruits in sugar with ascorbic acid	18	24	>24
Asparagus	15	24	>24
Carrots	18	24	>24
Cauliflower	15	24	>24
Green beans	15	24	>24
Beans (Lima)	18	>24	>24
Peas	18	>24	>24
Potatoes	24	>24	>24
Spinach	18	>24	>24

Dried organic products

Fruit are the principal imported dried produce and are easy to transport and to store. Dried vegetables are produced in low volumes for the local market but can be useful for e.g. soup mixes. The major risks with dried products are microbiological attack and physiological deterioration.

Physiological deterioration is caused by oxidation and enzymatic activity and leads to browning, loss of vitamins and the development of off-flavours. Dried organic products are particularly vulnerable to deterioration since chemical treatment with the normal conservation agent (SO2) is not allowed.

In order to minimise these risks a series of 'hurdles' are applied, as described in the following subsections. No organic substitutes are available which are as effective as (SO2) in conserving the colour of dried fruit. In these circumstarces a marketing campaign should be put into place at the point of sale to explain the reason for the brown colour of dried organic fruits, such as dried apricots.

Water content: Dry fruit products have a water content of 8 to 12% (in general) and dry vegetables, around 7%. Under these conditions, there are no microbiological problems during storage of the product. Semi-dried or soft products, which may be more highly priced in the market, have a water-content of between 20 and 30%; therefore microorganisms can develop during storage and refrigeration may be required.

Additives and processing aids: Permitted processing aids which help to retain quality of dried produce include:

— conservation agents and antioxidants: ascorbic acid, citric acid, tartaric acid, and salt;

— texturing agents: calcium chloride.

Ascorbic acid, citric acid, tartaric acid or lemon juice may be used for acidification. The resulting low pH limits the development of microorganisms and limits non-enzymatic browning. The product is treated by dipping in, or spraying with, acids or lemon juice.

Blanching : A brief period at high temperatures destroys most microorganisms present in the product; and inactivates enzymes which promote browning and degradation. Details of time, temperature, and solution vary according to the produce being treated; some examples are listed in Table 2.

Table 2. Recommendations for blanching fruits and vegetables.

Fruit	Process

Apricot	Steam for 5 min
Banana	Boiling water for 5 min
Date	Steam or boiling water
Litchi	Steam for 7 seconds, then soak in 5-10 % citric acid and 2 % salt
Mango, papaya	Hot water (56 °C) for 1 min
Pineapple	Steam for 10 min
Vegetable	Process
Cabbage	Boiling water for 3 min OR Steam for 5 min
Carrot	Boiling water for 4-6 min
French and green beans	Boiling water for 4-6 min
Gombo	Boiling water for 6 min
Spinach	Boiling water and salt (30 g/l), 3 min
Sweet potato	Boiling water for 5 min OR Steam for 8 min
Turnip	Boiling water for 4-6 min

Rapid Drying: Sun drying is the processing method most often used for organic fruits such as apricots, figs and bananas. However, the potential risk to quality and the difficulty of maintaining a high degree of sanitation is a problem. A rapid drying process also decreases the contact time between the product and oxygen. The drying conditions and recommended moisture content of the finished product are provided in Table 3.

Table 3. Drying conditions and moisture content of organic products

Fruit or vegetable	Drying temperature (°C)	moisture content of the finished product (%)	Expected storagelife (month)
Banana	55	12	6
Mango	55	14	6
Papaya	55	7-12	5
Onion	50-55	5	3-12

Tomato	55	6	6
Carrot	50-55	6	6-12
French beans	55	6	6

Water/air tight packaging : The type of packaging to be used for dried organic foods varies with expected storage conditions. The most common types are flexible packages with low permeability to oxygen and water vapour, and vacuum-packaging is common.

Pasteurisation after packaging : Certain processors pasteurise their products at 70°C after packaging to destroy microorganisms that could have contaminated the product after blanching. This treatment applies to soft fruits: apricots, plums.

Supplementary techniques for specific fruits: Dried dates are subject to considerable damage by pyralids (Ectomyelois ceratoniae). In order to avoid the use of chemical products (methyl bromide), two methods can be used to destroy this insect. Immediately after the harvest, dates can be placed into a dryer for 2 hours at 60°C in order to destroy larvae and eggs. Alternatively, the dates can be frozen at -40°C. Shock from cold temperatures seems to satisfactorily destroy the pyralid larvae and eggs. This procedure is more widely used.

Biological control has been tested. Among the techniques tried, one of them uses a bio-pesticide (bactospeine) and an insect (Habrobracon hebetor). When Bacillus thuringiensis is ingested, it produces the pesticide bactospeine in the organism. The use of B. thuringiensis is authorised by the EEC organic legislation. H. hebetor is a parasite to pyralids. When these two organisms are introduced into the storage room the rate of mortality of the pyralids reaches 76% after 6 days. Although effective, this method is not yet in commercial use.

After storage, dates are placed in a steam bath to stabilise the fruit's moisture level and to pasteurise it. Before packaging, the dates are sometimes coated with hot, pasteurised sugar syrup. This coating, which has a high sugar content of 70 to 80° Brix, acts as a barrier against contamination by microorganisms (due to the low absorption of water into the coating) and prevents the products from sticking together. In this manner, despite a high water content (24%), the development of microorganisms in date packaging seems to be rare, even among organic dates. However, to obtain these results, one must ensure that good hygienic conditions are applied during processing.

Dried organic grapes (raisins) are primarily dried in the sun. It is preferable to wash the grapes with water and oil before drying, and to spray them with organic sunflower oil before packaging. This helps stop the raisins from sticking together and acts as a barrier against the proliferation of microbes.

The by-products of dried fruits and vegetables are also used, incorporated into products like muesli. The incorporation of organic fruits and vegetables into other types of products is not as yet widely done, but products such as organic fruit yoghurt are coming on the market. To develop the market for dried fruits as a viable product in industry, the product must be available with consistent quality year-round and in sufficient volume to allow for storage.

Heat conservation process

These techniques use thermal treatments to conserve processed products by destroying or inactivating enzymes, and killing microorganisms. The products concerned are juices (principally from fruit), canned products (fruits and vegetables) and jams and purées.

Juice or Oil extraction: Juice is a simple and natural way to process fresh produce. It allows the preservation of the majority of nutritional qualities (vitamins, minerals), and it largely resolves the problem of storage. Juice could be considered as the ideal product for the organic market. The juices that are consumed in largest quantity (traditional and organic data, combined) are apple juice, citrus juices, pineapple juice and vegetable juices.

Among exotic fruit juices, the most common are from oranges, mangoes and guava. Carrot and tomato juices are important on the European market. There is probably a consumer view that juicing concentrates any residues in the product, which may explain the increasing popularity of pure organic fruit and vegetable juices. The market does not appear to be saturated as yet.

In addition, tropical fruits often offer a much higher source of vitamin C than fruits from temperate zones. This additional nutritional aspect is an advantage for tropical organic juices. It is preferable to opt for transport of juice in suitable bulk packages such as kegs to diminish transportation costs.

The juice can be repackaged in the country where it will be consumed. However, to ensure juice stability during handling, it is necessary to use high-temperature pasteurisation. This can alter the organoleptic and nutritional characteristics of the juice. To remedy this problem, it might be preferable to extract the juice at the location where the juice will be consumed.

This also allows the sale of 'fresh squeezed' or flash-pasteurised juices. This type of product is even more consistent with organic farming because little or no processing takes place and the final product is the closest thing to fresh. Fresh squeezed juice which has not undergone any type of heat treatment can be stored for only about 10 days under refrigeration. Flash-pasteurised juice (heat treatment of a few seconds at 70°C) has to stay in a cold room and can be stored for about 24 days. These products, the focus of ever-increasing consumer demand, correspond to the heightened interest in food products that are natural and healthy. It is therefore desirable to associate these qualities with the organic market.

Three factors are responsible for the successful long-term storage of juices:

— *The acidity of the juice* :The pH must be less than 4.2 in order to stop the development of microorganisms; ascorbic acid, citric acid or lemon juice can be added;

— *Thermal treatment* : A temperature which destroys microorganisms has to be used. Conditions depend on the characteristics of the final product: viscosity, packaging, type of juice, etc.;

— *The packaging* : Juice packaging has to be impermeable to gas and sometimes to light, and has to avoid micro-organism contamination.

Certain juices, especially those which are high in pulp content, may need to be clarified. Enzymatic treatments are authorised in Europe (part B of EC legislation), but filtration and decantation methods can be used instead. There is a growing market for organic oils, such as olive oil, which can be produced by simple physical methods from organic produce.

Canning : Canned produce must be prepared in such a way as to retain as closely as possible the characteristics of fresh produce. The market for canned products is restricted since consumer opinion relates canned products to products that are not wholesome, or are of poor quality.

Authorised additives and processing aids include:

— Antioxidants such as ascorbic acid, citric acid, and tartaric acid;

— Texture aids (calcium chloride).

Canning organic fruits and vegetables does not require any specific adjustments that are not already found in conventional canning methods. Fruit are usually canned in sugar syrup. This syrup is defined for 'natural products' (up to 16% sugar) or 'products with syrup' (over 20% sugar). Details of blanching recommendations (Table 4) and syrup compositions (Table 5) are given for canning a variety of common fruit and vegetables.

Table 4. Blanching recommendations for canned fruit and vegetables.

Fruit	Processing
Apricot	Boiling water for 1 min
Banana	Boiling water for a few min
Fig	Boiling water for 5 min
Grape	No blanching
Grapefruit	Steam for a few min
Vegetable	Process
Carrot	Boiling water for 2-4 min
Gombo	Boiling water for 2 min
Green pea	Boiling water for 8-10 min
Lima pea	Hot water (87-95 °C) for 3 min
Spinach	Hot water (71-77 °C) for 6 min
Sweet potato	Steam for 9-12 min

Table 5. Recommendations for syrup composition for canned fruit and vegetables.

Fruit	Syrup (g sugar per litre water)
Apricot	Natural: 80
	Syrup: 450
Banana	Syrup: 300 (+ 0.5% citric acid)

Cubic fruits	Syrup: 400
Fig	Natural: 80
	Syrup: 160-480
Grape	Syrup: 400
Grapefruit	Syrup: 400
Guava	Syrup: 650
Litchi	Syrup: 400-500 (+ citric acid)
Mango	Syrup: 400 (+0.25% citric acid)
Orange	Syrup: 500
Papaya	Syrup: 300-500 (+0.5% citric acid)
Pineapple	Syrup: 200-400
Vegetable	Syrup
Green peas	2 % salt, 4 % sugar
Carrot, French beans, Gombo, Lima beans, Spinach	2 % salt
Tomato, sweet pepper	2 % acidified salt
Sweet potato	Water

The time required for sterilisation of the cans depends on the characteristics of the product (viscosity, form, etc.), the size of the can and other factors. Oxygen in the produce can react with metals in the can material, so the cans need to have an internal protective coating to avoid corrosion. It is necessary to remove most of the oxygen from the produce by blanching and pre-heating.

Conservation with sugar: The goal of this processing technique is to arrive at a sugar content where the product is stable. This method is principally used for fruit. Durability in storage is determined by acidity; sugar content; and the type of packaging. There is a large variety of products: jams, jellies, syrups and fruit pastes. Importing countries may impose standards and definitions for these products, specifying the minimum quantity of fruits or fruit juice that may be used, as well as the optional ingredients.

Conservation with sugar does not require any particular modifications for organic farming; although where possible organic sugar sources will be preferred. The process consists of mixing sugar (or sugar syrup) with pulp of fruits (or fruit juice). The mixture is cooked to remove excess water. Pectin solution is added. When the desired end point is reached (65° Brix for jams and jellies, 68° Brix for preserves), the proper amount of acid is added. Once the pH has been adjusted to 3.1 to 3.4, the mix is ready for packaging.

The low pH and water activity of the product limit bacterial development. However the low solids content (65-70%) is not a guarantee against the growth of certain microorganisms. To ensure total protection, post-filling sterilisation could be carried out.

The ingredients and processing acids required for these products are:

— preserving agents such as lactic acid, citric acid, calcium citrate, tartaric acid, sodium and potassium tartrate;

— thickening and gelling agents such as pectin, flour from carob beans, guar gum, adragante gum, arabic gum, alginic acid, sodium and potassium alginate, agar-agar, carragheenan;

— processing aids such as calcium chloride, coatings (Carnauba and bees wax), and calcium carbonate.

Purées and pulps are used in products for children. Some major manufacturers of juice for babies have oriented their operations towards products without pesticide residues or preservatives. This means that organic materials were used to make the products. For the moment, only banana pulp is well developed. However, as with juice, the nutritional aspect of tropical fruits could stimulate a market for other types of pulps and purées.

Jams also offer a potential market for pulps and purées. As with juice, these are products that have been only slightly processed, are healthy, and that keep the nutritional properties of the original fruit or vegetable and they correspond to the idea of health that is associated with organic products. Jams are primarily sold in fine markets or specialised shops. Some blanching details for stewed fruit are listed in Table 6.

Table 6. Blanching conditions for stewed fruit.

Fruit	Processing details
Apricot	Hot water (80-85°C) for 2 min
Mango	Steam for 2 min
Orange	Boiling water for 5 min

Fermentation

Fermentation is a chemical change brought about by enzymes, bacteria or microorganisms. The chemical changes are acidification, oxidation of nitrogenous organic compounds and decomposition of sugars and starches (alcoholic fermentation). These last fermentation options result in wines and other alcoholic beverages. Organic processing methods for alcoholic beverages are not addressed here.

There are many other important fermentation processes and products. They principally concern vegetables: sauerkraut, pickled vegetables, fermented soya (shoyu, tempeh), and fermented roots or tuber

Fermentation is achieved by:

— salt, without water ;

— brine: salted solution, sometimes with vinegar;

— the inoculation of specific microbes (shoyu, tempeh).

— water (cassava).

Preparations of microorganisms and enzymes commonly used in food processing may be used, with the exception of genetically engineered microorganisms and their products. Microorganisms grown on organic cultures should be used if available.

Sauerkraut: Cabbages are cut and put in fermentation tanks with salt. The salt is mixed with the shredded cabbage and lactic fermentation causes the conversion into sauerkraut. When the acidity is sufficient the fermentation is stopped by pasteurisation. The process does not require specific additives or processing aids. Canned sauerkraut is easy to use and is not as subject to deterioration and spoilage as bulk sauerkraut. Sauerkraut used for canning is ordinarily not cured as long as that sold in bulk.

Pickles: The cucumber is one of the most important vegetables used for pickles. Sometimes vegetables undergo a preliminary fermentation with salt. They are soaked in several changes of cold water until practically free of salt, followed by many hours in hot water (45-65°C). Calcium chloride has a hardening effect and can be added. Afterward, vegetables are placed in vinegar. Pickles are put into heavily lacquered cans with brine or vinegar. The cans undergo thermal treatment and are sealed.

Other fermentation: The fermentation of soya involves the use of microorganisms: Saccharomyces, Torulopsis, Pediococcus or Rhizopus. These elements are authorised by EC Regulation. The fermentation of cassava (or other roots and tubers) is a natural fermentation that needs no additives.

Fermentation is a viable process for enhancing the microbiological safety and storage of foods, especially in areas such as the tropics. It conforms to organic legislation. Many types of new fermented products can be created such as vegetable cheeses and yoghurt.

Organic Legislation

The development of organic farming increases the risk of fraud and of unfair competition. In order to protect consumers and producers, many countries have put in place requirements that standardise products at the national and the international level. An increasing number of countries have standards and regulations in place. Details for major importing countries can be found at http://www.organic-research.com/Laws&Regs/legislation.htm. These regulations are in addition to the requirements of any national certifying or registering body in the country of production; and should be read in conjunction with the Codex Alimentarius guidelines and IFOAM draft basic standards.

One of the first pieces of organic legislation was the European Union's "Council Regulation (EEC) No. 2092/91" from June 24, 1991. The full regulation and its amendments are available on http://europa.eu.int/eur-lex/en/lif/dat/1991/en_391R2092.html. It defines the production methods for organic farming: rules for labelling, production, quality control systems, and imports from the developing world.

The requirements for organic farming are applied in an identical fashion in all the EU-countries, although not all products may be authorised for cultivation in every country. In addition, in each country of the EU, organic products can exist with even stricter requirements. These products must therefore undergo a double certification:

European certification and certification tied to the particular product.

For countries wishing to export organic products to the EU, there are basically two options. Suggestions on how to proceed if a country should wish to attempt inclusion on this list can be found in; but the process is long and difficult. The more practical (but still expensive) alternative is to apply for an individual permit to import; certification will be required by both the producer and exporter, which may occur through an official accrediting body of the importing country, or through a local certification body accredited by the importing country. Further details are available in, pp. 10-15.

Switzerland has its own law concerning organic farming, 'L'ordonnance sur l'agriculture biologique', 22 Sep 1997. Turkey's legislation on organic agriculture is described in. In Japan, the Japanese Agricultural Standards (JAS) legislation for organic agriculture was implemented in April 2000. This legislation replaces earlier voluntary guidelines which covered a range of 'green' agricultural practices; there is now a clear definition of organic produce and labelling rules.

As far as the USA is concerned, the US National Organic Standards were completed in December 2000 and will come into full effect by October 2002. Descriptions of entry requirements for organic produce are described in.

Storage and processing of organic agricultural products: Regulation may be used only where it has been shown that such ingredients are not produced in sufficient quantity in the community or cannot be imported from their country according to the legislation. The authorisation is accorded for a period of 3 months.

Labelling: Labelling and advertising organic products must be closely regulated in order to avoid fraud and to protect the consumer. Labelling and advertising a product may refer to organic production methods only where:

— The product was produced in accordance with the rules ;

— The product was produced or imported by an operator who is subject to the inspection measures l;

— All the ingredients of agricultural origin have been provisionally authorised by a Member State;

— The product or its ingredients have not been subjected to treatment involving the use of ionising radiation.

Labelling must list the following indications:

— The name and/or code number of the inspection authority or body which has carried out the most recent inspection;

— A clear statement that the words relate to a method of agricultural production.

The other elements listed depend on the percentage of organic products included in the final product.

In Swiss legislation, the same labelling norms are applied. In Turkey, the same rules apply for labelling, but other indications should appear:

— The content of non-organic agricultural ingredients;

— The name and the content of the non-agricultural ingredients.

For the USA, the National Organic Programme defines very strict rules concerning labelling of organic products. Any agricultural product that is sold, labelled, or represented as organic shall comprise 100 percent of the total weight of the finished product, excluding water and salt, except that not more than five percent of the total weight of the finished product may consist of one or more of non-organic allowed ingredients.

Any agricultural product that is sold, labelled or represented as made with certain organic ingredients on the main display panel shall comprise at least 50 percent, but less than 95 percent, of the total weight of the finished product of non-organic allowed ingredients. Labelling standards are rigorously detailed in the Act (Subpart C).

PEST CONTROL

Pest and diseases

There are not a lot of approved organic post-harvest treatments for pests and diseases. The guidelines usually emphasise the need to minimise pest and disease pressure before harvest. Hot-water (45-55°C) immersion, steam and forced hot-air treatments are sometimes used as organic control methods after harvest. Most pathogenic microorganisms are destroyed within a few minutes of hot-water immersion.

The disadvantages are the risk of product damage: re-humidification of the surface increasing the risk of degradation of the fruit or vegetable; acceleration of ripening; damage to the colour and firmness of the flesh. It is necessary to maintain a very precise temperature (to within a half degree) and treat for a precise time.

For pathogenic microorganisms, there are antagonistic floras. For example, Bacillus subtilis slows the development of several moulds. In order to be effective, it is necessary that the antagonistic flora are stable, can develop under storage conditions, and are tolerated by the consumer. These techniques are not yet in commercial use for stored fruits and vegetables.

8

Environmental Impact of Organic Agriculture

Environmental impact is a fundamental issue to be taken into consideration for a comprehensive evaluation of the role played by organic farming. Environmental, health and social aspects are strictly related with the particular organic approach to farming, but their evaluation may require great efforts for data collection and analysis.

Currently, the most relevant environmental indicator system has been originally presented OECD in 1997, and recently modified and updated. It is developed with regard to environmental and resource use effects in order to enable the analysis of country-specific situations, to evaluate environmental policies and to measure environmental quality. OECD provides also a set indicators specifically targeted to the agricultural sector; it is based on the Driving Force—State Response (DSR) approach and it is based on solid methodology concerning the environment and natural resources measurement.

There is a general agreement in literature that the DSR framework improves substantially existing approaches to environmental indicator concepts, and can be considered as a reference point for environmental analysis. OECD approach has been recently taken into consideration by

EU Commission for statistical data collection purposes and by FAO for a critical discussion and review of relevant agri-environmental indicators systems.

Environmental indicators aim:

— to help monitor and assess agri-environmental policies and programmes, and to

— provide contextual information for rural development in general;

— to identify environmental issues related to the investigate Country's agriculture;

— to help target programmes that address agri-environmental issues;

— to understand the linkages between agricultural practices and the environment.

As indicated in COM144, the main criteria for choosing agri-environmental indicators are:

— *relevance* -address the key environmental issues

— *responsiveness* -change sufficiently quickly in response to action

— *analytical soundness* - based on sound science

— *measurability* - feasible in terms of current or planned data availability

— *ease of interpretation* -communicate essential information in a way that is unambiguous and easy to understand;

— *cost effectiveness*-costs in proportion to the value of information derived

The OECD set of environmental indicators for agricultural sector contains several sub-categories assigned to each of the three DSR components:

Driving forces:

— Environment

— Economy and society

— Farm inputs and outputs

State:

— Ecosystem

— Natural resources

— Health and welfare

Response:

— Consumer reaction

— Agro-food chain responses

— Farmers behaviour

— Government policies

Stolze et al have used a qualitative approach for the study of the environmental effects of organic farming in Europe within the EU project "Organic Farming and the CAP". The analysis evaluates organic farming environmental performances in comparison with conventional farming, that represents therefore an appropriate reference system. A general theoretical problem is the individuation of the appropriate farming system to use for comparison:

— typical farming system;

— best practice management;

— management plus specific measures for environmental protection.

This means for example that different environmental performance of conventional farming may be considered, like integrated farming, high-input conventional farming, low-input farming. Analogously different variation of organic farming may be taken into consideration, like "standard" organic, best practice organic farming, and highly environmental friendly organic farming (e.g. providing exclusive areas of "pure nature").

The choice of concentrating the analysis on the comparison of organic and conventional environmental performance allows to simplify and adapt the original OECD indicators set. Concerning the Driving forces, only farm input and output can be considered, as environmental, economic and social frameworks are not a defined characteristic of the farming system, nor does a farming system affect these factors.

The approach of Stolze et al is a step by step qualitative assessment for data aggregation:

— firstly each parameter for a given environmental indicator is analysed on the basis of available data and literature;

— secondly the results of the studies are reviewed and commented;

— finally the parameter results are aggregated by indicator.

The aggregation procedure is based on a final qualitative assessment of the organic farming performance with respect to conventional farming, according to the scheme of Table 1.

Table 1. Assessment scale used for indicator evaluation

Scale Organic		farming performs...
++	=	much better
+	=	better
0	=	the same
-	=	worse

	=	much worse ...than conventional farming

Life Cycle Assessment (LCA)

One of the most prominent empirical approach for the measurement of environmental impact of an industry, sector or product is Life Cycle Assessment (LCA). LCA is a methodology used for analysing and assessing the environmental loads and potential environmental impacts of a material, product, or service throughout its entire life cycle, from raw materials extraction and processing, through manufacturing, transport, use and final disposal. Early forms of LCA were used in the United States in the late 1960s for defining corporate environmental strategy, and later in the 1970s by government agencies as an aid for developing public policy. In the late 1990s, LCA emerged as a worldwide environmental management tool in the form of the ISO 14040 series.

The methodology for LCA in the food sector has been developed during the last decade and progress in terms of methodological robustness and data availability has been demonstrated—among others—at a series of LCA-food conferences organised in Brussels and Gothenburg. The interpretation of the results from previous phases of the study is made in relation to the original objectives established initially.

Life Cycle Assessment Tools

The LCA tool consists of a number of discrete stages, and are the follow:

Scope definition

The first phase of the LCA study involves defining the purpose of the study, its scope, data quality goals, and functional unit. The project sponsor generally shapes the purpose and scope, and it is important that they are clearly defined to avoid any subsequent misunderstandings about the wider application of the results. LCA objectives can be classified broadly into system improvement studies, in which the goal is to identify opportunities for reducing the environmental effects of an existing system or process, and comparative studies, in which the intent is to select an optimal product of process from a number of predetermined alternatives. Scope definition involves specifying system boundaries, functional unit, allocation assumptions, inventory parameters, and impact categories that will be used.

Life cycle inventory (LCI)

Detailed life cycle inventory (LCI) analysis, with compilation of data both about energy and resource use and on emissions to the environment, throughout the life cycle. This phase involves the quantification of environmentally relevant material and energy flows of a system using various sources of data. Essentially, an accounting of system inputs and outputs is performed. The data used may come from a variety of sources, including direct measurements, theoretical material and energy balances, and statistics from databases and publications.

Data quality can often be a problem in LCA's, with complete absence of data perhaps being more of a difficulty than data quality. Definition of an appropriate functional unit is fundamental to the credibility of an LCA. The functional unit is the unit of analysis for the study, and it provides a basis for comparison if more than one alternative is being studied. It should be defined in terms of the services provided by any product, process, or activity under analysis.

An assessment of the potential impacts

An assessment of the potential impacts associated with the identified forms of resource use and environmental emissions. It analyses and compares the environmental burdens associated with the material and energy flows determined in the previous phase. The conventional approach is to classify the inventory flows into specific impact categories. Normalisation and weighting (or valuation) of the impacts is also included in this stage. If necessary, the individual impacts can then be aggregated into a single composite environmental index.

Interpretation

The interpretation or Improvement Assessment utilises the results of the preceding stages to meet the specified objectives. Typically this phase will generate a decision or plan of action. For diagnostic LCAs, data is used to identify critical segments or "hot spots" in the life cycle, which contributes disproportionately to the total system environmental impact. These problem areas can then be eliminated or reduced through system modifications. In the case of comparative LCAs, the competing system life cycles are ranked based on environmental performance and the optimal alternative is selected.

Applications

LCA is applied in different perspectives and different areas and is mostly used in the following contexts:

Policy and practice

There is an increasing interest in using LCA to underpin public sector decisions on policy and practice. For example, as the basis for European environmental legislation including Integrated Pollution Prevention and Control (IPPC) and Integrated Product Policy.

Product design

Design for the environment is an approach that uses a simplified life cycle approach to enable a product developer to obtain a quick assessment of the implications of.

Process selection

Within the processing, currently processes are selected and plants designed by considering only the immediate process. However, local optimization of a process can lead to overall reduction of environmental performance, typically by transferring environmental burdens to other points in the life cycle. Design of the whole life cycle rather that just one process has yet to be properly developed and accepted.

Supply chain management

A purchaser at one point in the life cycle of a product may have greatest scope to improve the product life cycle by selecting suppliers with better environmental performance.

For the implementation of Life Cycle Analysis different software are available. They include different methods of calculation and different databases for different uses and applications, depending on the original objectives requirements.

Steps

An LCA study consists of four steps:

Goal and scope of the study

In the goal and scope the most important choices are described, such as:

— The reason for executing the LCA, and the questions, which need to be answered.

— A precise definition of the product, its life cycle, and the function it fulfils. In case products are to be compared, a comparison basis is defined (functional unit).

— A description of the system boundaries.

— A description of the way allocation problems will be dealt with.

— Data and data quality requirements.

— Assumptions and limitations.

— The requirements regarding the LCIA procedure, and the subsequent interpretation to be used.

— The intended audiences and the way the results will be communicated. If applicable, the way a peer review will be made.

— The type and format of the report required for the study.

Goal: It is obvious any LCA study should have goal. However, in ISO there are some particular requirements to the goal definition: The application and intended audiences shall be described unambiguously. This is important, as a study that aims to provide data that is applied internally can be quite differently structured than a study that aims at making public comparisons between two products.

Scope: A particularly important issue in product comparisons is the functional unit or comparison basis. In many cases, one cannot simply compare product A and B, as they may have different performance characteristics. For example, a milk carton can be used only once, while a returnable milk bottle can be used ten or more times. If the purpose of the LCA is to compare milk-packaging systems, one cannot compare one milk carton with one bottle. A much better approach is to compare two ways of packaging and delivering 1000 litres of milk. In that case one would compare 1000 milk cartons with about 100 bottles and 900 washings (assuming 9 return trips for each bottle).

System Boundaries: It is helpful to draw a diagram of the system and to identify the boundaries in this diagram.

Important choices in this area are:

— Will the production and disposal of capital goods (trucks, injection moulding machines etc) be included?

As in energy analysis, one can distinguish three orders:

— First order: only the production of materials and transport are included (this is rarely used in LCA).

— Second order: All processes during the life cycle are included, but the capital goods are left out.

— Third order: Now the capital goods are included. Usually the capital goods are only modelled in a first order mode, so only the production of the materials needed to produce the capital goods are included.

— What is the boundary with nature?

Allocation: Many processes usually perform more than one function or output. The environmental load of that process needs to be allocated over the different functions and outputs. There are different ways to make such an allocation. ISO recommends the following procedure in order to deal with allocation issues:

— Avoid allocation, by splitting the process in such a way that it can be described as two separate processes that each has a single output. Often this is not possible, for example, wooden planks and sawdust are both an economic outputs of a sawmill, but one cannot split the sawing process into a part that is responsible for the sawdust and one that is responsible for the planks.

— Another way to avoid allocation is to extend the system boundaries and by including processes that would be needed to make a similar output. For example, if a usable quantity of steam, produced as a by-product, is used in such a way that it avoids the production of steam by more conventional means, one may subtract the environmental load of the avoided steam production. A practical problem is often that it is not always easy to say how the steam would be produced alternatively.

— If it is not possible to avoid allocation in either way, the ISO standard suggests allocating the environmental load based on a physical causality, such as mass or energy content of the outputs. For example if the sawdust represents 40% of the mass, one can allocate 40% of the environmental load to sawdust. In the case of allocating steam.

— If this procedure cannot be applied, ISO suggests using an socioeconomic allocation basis, such as the economic value.

Model of the product life cycle

Making a model of the product life cycle with all the environmental inflows and outflows requires a data collection. This latter step is usually referred to as the life cycle inventory (LCI) stage.

Inventory: The most demanding task in performing LCAs is data collection. However much data is available in the database you will usually find that at least a few processes or materials are not available, or the available data is not representative. Depending on the time and budget you have available, there are a number of strategies to collect such data. It is useful to distinguish two types of data:

— Foreground data

— Background data

Foreground data refers to very specific data you need to model your system. It is typically data that describes a particular product system and particular specialised production system. Background data is data for generic materials, energy, transport and waste management systems. This is typically data you can find in databases and literature.

Life Cycle Impact Assessment (LCIA)

The study stage of Life Cycle Assessment intends to understanding the environmental relevance of all inflows and outflows. Most LCA experts do not develop impact assessment methodologies, they prefer the selection of what has been publish. Like in the inventory stage, also in impact assessment the Goal and Scope definition is the most important source of guidance for the selection of the method and the impact categories.

The most important choice you make is the desired aggregation level of the results. This usually depends on the way you would like to address your audience, and the ability of your audience to understand detailed results. The ISO 14040 standard defines an LCA as a compilation and evaluation of the inputs and outputs and the potential environmental impacts of a product system through its life cycle. In this definition, it is clear that impact assessment is an integral part of LCA. The ISO 14040 standard defines the impact assessment phase as the "phase of a LCA assigned to the study and evaluation of the possible environmental impact caused by the system-product under study and has the aim to identify and underline the entities of the modifications generated by resources' consumptions and by the discharges in the environment, included in the inventory". In brief, Life Cycle Impact Assessment is defined as the phase in the LCA aimed at understanding and evaluating the magnitude and significance of the potential environmental impacts of a product system. It is considered as the transition phase between the data collection and analysis to the study of the environmental impacts.

The impact assessment methods themselves are described in ISO 14042. In this standard, a distinction is made between:

— Mandatory elements, such as classification and characterisation

— Optional elements, such as normalisation, ranking, grouping and weighting.

This means that according to ISO, every LCA must at least include classification and characterisation. If such procedures are not applied, one may only refer to the study as a life cycle inventory (LCI). An important distinction is made between internal and external applications. If results are intended to compare (competing) products and they are to be presented to the public, weighting may not be used.

An important step is the selection of the appropriate impact categories. The choice is guided by the goal of the study. An important help in the process of selecting impact categories is the definition of so called endpoints. Endpoints are to be understood as issues of environmental concern, like human health, extinction of species, availability of resources for future generation etc. ISO does not recommend using certain endpoints 4, but requires a careful selection and definition of endpoints first. Subsequently, impact categories can be selected, as long as the environmental model that links the impact category to the endpoint is clearly described. It is not necessary to describe this link quantitatively.

Classification: The inventory result of an LCA usually contains hundreds of different emissions and resource extraction parameters. Once the relevant impact categories are determined, these LCI results must be assigned to these impact categories. For example CO_2 and CH_4 are both assigned to the impact category "Global warming", while SO_2 and NH_3 are both assigned to an impact category acidification. It is possible to assign emissions to more than one impact category at the same time; for example SO_2, may also be assigned to an impact category like Human health, or Respiratory diseases.

Characterisation: Once the impact categories are defined and the LCI results are assigned to these impact categories, it is necessary to define characterisation factors. These factors should reflect the relative contribution of an LCI result to the impact category indicator result. For example, on a time scale of 100 years the contribution of 1 kg CH_4 to global warming is 42 times as high as the emission of 1 kg CO_2. This means that if the characterisation factor of CO_2 is 1, the characterisation factor of CH_4 is 42. Thus, the impact category indicator result for global warming can be calculated by multiplying the LCI result with the characterisation factor.

Normalisation: Normalisation is a procedure needed to show to what extent an impact category has a significant contribution to the overall environmental problem. This is done by dividing the impact category indicators by a "Normal" value. There are different ways to determine the

"Normal" value. The most common procedure is to determine the impact category indicators for a region during a year and, if desired, divide this result by the number of inhabitants in that area.

Normalisation serves two purposes:

— Impact categories that contribute only a very small amount compared to other impact categories can be left out of consideration, thus reducing the number of issues that need to be evaluated.

— The normalised results show the order of magnitude of the environmental problems generated by the products life cycle, compared to the total environmental loads in Europe.

Damage assessment / Ranking: The methods presented here will all still be difficult to interpret, as there is a wide range of impact category indicators. To simplify interpretation further, a grouping procedure can be used in the Ecoindicator 99 and the EPS2000 methodology. In these methods, the category indicators are defined close to one of the three endpoints to achieve an optimum environmental relevance. The impact category indicators that refer to the same endpoint are all defined in such a way that the unit of the indicator result is the same. This allows addition of the indicator results per group. This means that the indicator results can be presented as three indicators at endpoint level without any subjective weighting. Interpreting three instead of a multiple set of indicators is much easier.

Weighting: Weighting is the most controversial and most difficult step in life cycle impact assessment, especially for midpoint methods. Several solutions have been proposed to solve or simplify the weighting problem:

— Use a panel that assesses the impact category and proposes default weights. There are several problems in this approach:

  — It is very difficult to explain to a panel the meaning of the impact category indicators. They are too abstract ("CO2 equivalency" or "proton release").

  — In a Midpoint approach, the number of indicators to be assessed is usually rather large (10 to 15).

  — Panels tend to give a very small range of weights. This is called framing in social sciences. This is a problem in both endpoint and midpoint methods.

— Distance to target. If it is possible to set a target for each impact category and its target can be used to derive at a weighting factor. If the difference is high, the weight is high. The Ecopoint method uses targets set by the Swiss government, the Eco-indicator 95 method uses targets that reflect to necessary reduction to lower the damage to a certain level that is the same for all impact categories (this can also be interpreted as a damage approach). Also this approach has some difficulties:

  — In the case policy targets are used, it is not clear if all targets are equally important.

  — policy targets are usually formed as a compromise between interest groups, and need not to reflect the "real" need to reduce environmental impacts.

  — In case scientific targets are used, different types of damages need to be weighted.

— Monetarisation: In EPS2000 all damages are expressed in the same monetary unit: Environmental Load Units comparable to Euros. In the methodology, the assumption is made that these different types of costs can be added. This can be interpreted as a weighting step in which the weighting factors for these different types of costs equals one.

In the Eco-indicator 99 methodology, the weighting problem was the starting point of the development. Some of the problems associated with the weighting have been reduced or solved, but the weighting step will always remain difficult. An interesting approach has been developed by using a weighting triangle. This triangle can be used to present the weighting problem on a case-by-case basis to stakeholders. It can be used to take a decision without actually knowing the weights.

Life Cycle Interpretation

Life Cycle Interpretation and Improvement is defined by ISO 14040 norm as "Phase of a LCA in which the inventory results and/or of the impact analysis, are elaborated correspondingly to the study objective and in such to reach conclusions and recommendations". The interpretation phase is the conclusive phase of the LCA that seeks to propose or suggest basic changes to reduce the environmental impact.

Software for LCA Analysis

Different software for the LCA analysis are available on the market. One of the most versatile is SimaPro. SimaPro 5 offers a wide number of alternative methods to carry out this phase of the study:

CML 1992

CML 2 baseline 2000

Ecoindicator 95

Ecoindicator 99

Ecopoints 97

EDIP/UMIP 96

EPS 2000

Ecoindicator 99

The Eco-indicator 99 is a state of the art impact assessment method for LCA, with many conceptual breakthroughs. The method is also the basis for the calculation of eco-indicator scores for materials and processes. These scores can be used as a user friendly design for environment tool for designers and product managers to improve products.

Weighting is a controversial step in impact assessment. Taking weighting as the starting point of the Eco-indicator 99. This is in contrast with the "bottom up" approach found in the traditional theme oriented methods, where a panel can be confronted with ten or more abstract environmental themes to weight. New damage models and 3 perspectives were developed to complete the method.

Damage categories: The need to simplify the weighting procedure needed to define the "Eco" wanted to "indicate" first. The following damage categories were defined:

— Damage to Human Health

— Damage to Ecosystem Quality

— Damage to Resources

New damage models were developed that link inventory results into three damage categories.

Weighting results: The result of the panel procedure is that Human Health and Ecosystem Quality are considered to be of almost equal importance, while Resources are considered to be half as important.

Human Health: Damages to human health are expressed in Disability Adjusted Life Years or DALY's. This method, developed by Murray, is used by WHO and WorldBank. An important element is a scale that rates the different disability levels. Damage models were developed for respiratory and carcinogenic effects, the effects of climate change, ozone layer depletion and ionising radiation.

In these models four steps are used:

1. Fate analysis, linking an emission (expressed as mass) to a temporary change in concentration.
2. Exposure analysis, linking this temporary concentration change to a dose.
3. Effect analysis, linking the dose to a number of health effects, such as occurrence and type of cancers.
4. Damage analysis, links health effects to DALYs, using estimates of the number of Years Lived Disabled (YLD) and Years of Life Lost (YLL), (following).

Ecosystem Quality: Damage to ecosystem quality are expressed as percentage of species disappeared in a certain area, due to the environmental load (Potentially Disappeared fraction or PDF). The PDF is then multiplied by the area size and the time period to obtain the damage. The damage category Ecosystem Quality is unfortunately not as homogeneous as the definition of Human Health. It consists of Ecotoxicity, Acidification and Eutrophication, Land use and land transformation.

Resources: Damages to Resources, minerals and fossil fuels, are expressed as surplus energy for the future mining of resources.

9

Socio-economic Impact of Organic Agriculture

The wide range of country-specific environment for organic farming requires an analysis of the regulatory framework and policy context that may be encountered in order to properly evaluate the possible effects of organic farming uptake. A preliminary step is the individuation of the exact definition and terminology for organic farming assumed as official in each country; in fact, often different synonymous for organic farming are used, like ecological farming, biological farming etc. Following Lampkin et al, the basic scheme for the analysis of the regulatory framework for organic farming policies is as follows:

- General (i.e. conventional and organic) agro-environmental policy system, with a description of the role played by organic farming policies;
- Support schemes specific for organic farming (national and/or regional level);
- Other agro-environmental and rural development support schemes that may indirectly affect organic farming;
- Marketing and processing measures for organic products and foreign trade measures;

— Advisory services, certification, education and training, research and development; Data collection can be carried out by local specialists, that should ideally be native speakers and recognised expert in organic farming. In order to reach a comprehensive description of the regulatory framework, two approaches may be used jointly:

— direct contact with key informants through phone, postal or personal interviews;

— literature review of the topic.

Concerning the first approach, a standardised questionnaire shall be defined including a "core" set of basic information that must be collected for each country, and a secondary set of information that shall be included if available with a reasonable level of effort.

Agricultural Sector Policies

A general overview of the policies for the agricultural sector in the countries where the analysis will be performed is essential to understand the relative weight given to organic farming and to assess if and to what extent agriculture policies are actually beneficial for the organic sector. A preliminary step is a short description of the agricultural sector in terms of main agricultural productions, number of farmers, average farm size, foreign trade balance for agro-food sector.

The FAO legal database can be a useful basis for this task. The quantitative description may be integrated by a qualitative SWOT analysis. The description of agricultural sector should be accompanied by a short description of the evolution of the agricultural policy system in the investigated countries covering the key elements and relevant turning points over the last decades.

This step of the analysis requires a description of the basic framework of agricultural policies, aiming to understand the overall public support for the agricultural sector. If relevant, for each country the issues to be investigated are in particular:

Protectionist policies for agricultural incomes, like tariff and/or non tariff barriers to trade, export Subsidies, price integration, direct income support, etc;

— Aids for investments and purchase of inputs;

— Agro-environmental support schemes;

— Technical support and extension services;

— Any other relevant policy aspects.

A further issue to be considered is the eventual country membership to international trade partnerships that might be relevant for agricultural trades, taking into consideration:

— description of the general aim of the partnership;

— list of other member states;

— description of the most influenced agricultural production;

— possible direct or indirect effects on organic products trades.

A description of the national and local authorities for agricultural policies should be included, with a short description of the role played by each actor in the agricultural policy system. At this stage it is essential to individuate clearly all the actors involved directly or indirectly in the definition and implementation of the organic farming policies both at a national and local level.

Organic Farming

The description of the specific policies and support schemes for organic farming should consider in particular the general aim of the regulations, and the specific areas of interest. All the relevant regulations should be listed, identifying when they have been issued, and the specific areas of concern. Issues covered by these regulation regard the following areas:

— Conversion to and maintenance of organic farming methods

— reduction of pesticides and fertilisers,

— weed and plant control,

— production of forage for livestock productions,

— livestock productions,

— agencies/regulation of the certification system

— reliability of the certification and accreditation system.

For each measure a short description of the implementation should be provided, giving a quantitative description in terms of numbers of hectares, livestock units, and farms that have been interested by the organic schemes. Such information may be accompanied by a description of the total public expenditure per implemented measure.

The description of the measures for organic farming should be integrated by that of the eligibility conditions of applicants, providing a description of the following aspects:

— eligible farmers typology: consider if partial and staged conversion is allowed, maximum time period allowed for conversion;

— eventual certification requirements: consider if official certification from a certification body is required to be eligible for aids, if there are differences between crop and livestock certification, and if there is a minimum time period of maintenance of the organic scheme;

— farm size: considers if there are maximum and minimum size requirements;

— eligible crops: considers if there are crops excluded from support;

— livestock density and livestock number: considers eventual requirements in terms of stocking rate and LU.

Economic support for organic farming is crucial to evaluate the actual attractiveness of organic farming support measures. Information is required concerning the kind of support available for organic farming, and the criteria used for the support definition. In particular it should be pointed out if support is justified in terms of income loss due to conversion or if it incorporates also additional incentives, like possible payments to the processing small and medium enterprises sector.

Particularly relevant is the description of the eventual payment diversification for crop, livestock typology and farm type and location. A quantitative description of payment level should be provided according to the scheme adopted for payments in the country analysed. Concerning payments, it should be considered if they are diversified according to the following cases:

— converting or continuing organic farms: helps to recognise if conversion costs are specifically taken into account;

— period of conversion: considers if payments are stable through time or if they reduce progressively;

— farm size: considers if payments are reducing as farm area increases.

A quantification of the public expenditure on organic farming should refer to the main areas of funding, and possibly considers payment per farm and/or crop types. These figures would help the assessment of the actual effectiveness of organic policies, and constitute the basis for a comparison with benefits arising from organic farming diffusion, that will be possibly measured in the further steps of the analysis.

Rural Development Schemes

Other agroenvironmental and rural development measures, even if not explicitly focused on organic farming, may produce side effects that may influence organic farming both positively and negatively. Possible positive effects may be those concerning extensification programmes, specific payments for less favoured areas, specific policies for labour intensive farm structures, policies for high quality food products etc. A qualitative assessment of the conditions and motivation which may favourite organic farming may help in integrate the quantitative description of the organic payments.

Marketing and Processing Measures for Organic Products

Market may play a fundamental role for the competitiveness of organic products, specially in conditions where public support is insufficient to cover the gap of economic performance with respect to conventional products: premium prices, product differentiation, higher quality and environmental friendly image are some of the usual "plus" of organic products. In situations where financial support for conversion and maintenance of organic farming is low, organic farmers may rely only on consumers willingness to pay higher prices in order to obtain a compensation for lower yields or higher costs that may arise due to the organic practice.

Appropriate demand pull policies for organic products may contribute to reduce the overall public expenditure required for an increase of organic production as they allow for lower supply subsidies, and can represent therefore a feasible solution for developing countries where budget restriction represent an obstacle for supporting organic farming.

A description of the measures available for market development should investigate how they integrate with policy-driven supply increase, and if they lead to market distortion or unfair competition. Also, it is necessary to indicate the organisation playing the main role in organic market development, individuating if it is public or private, and if it operates on a national or international level.

Certification Standards and Advisory Services

Standards for organic farming are intended to promote consumers confidence and prevent the undermining of the market through fraudulent trading. They are based on legislation and definitions that may be set on a international or national level.

Concerning the international level it is necessary to consider if IFOAM or CODEX Alimentarius definition of organic farming and food are adopted in the country under investigation. These basic standards can be considered as minimum level standards that are internationally accepted and provide a basis for developing more detailed standards at a national level. National legislation and regulation for organic standards and certification may be based on international requirements: a description of how and if it is integrated nationally is therefore necessary. The role and activity of inspection and certification bodies in particular should be described, identifying their position in the national and international context on the matter.

Advisory services, training and education, research and development This step aims to identify the kind of advisory, extension and information services available, providing (if available) a description of the number of farmers using such services and indicating if the services offered are integrated with the main agricultural extension services or if there are specific services for organic farming. A description of the main types of advisory and extension services is necessary, together with that of the main actors offering such services.

Analogue description should be provided for training and education services: range of training types available, kind of institutions offering the services. Concerning the research and development activities for organic farming it should be described the institutions involved and possible cooperation among them both on a national and international level. Besides, a description of the main projects and research activities should be provided, with a short indication of the main themes considered.

Adoption of Organic Farming

The difference between organic and conventional land-use patterns; both influence the quantity and composition of agricultural output due to the conversion to organic schemes. Again, a wide range of factors that can directly or indirectly affect land-use pattern changes should be taken into account. These are in some cases difficult to account for and often depend on country-specific characteristics.

In particular, attention should be paid to specific rotation schemes adopted by organic farming, which cause material differences in areas harvested under different crops. Furthermore, organic farming manuring is often heavily based on livestock production, therefore causing a tendency to reallocate farm activity in order to balance livestock and crop production according to the proper organic management of the farm.

The basic variables influencing output variation may also have a direct influence on public expenditure. Different land-use patterns may modify direct payment expenditure through a reallocation of payments given on a per crop or type of animal production; impacts on public expenditure may arise also referring to the management of foreign trade of agricultural commodities.

Organic conversion might also cause several spill-over effects on public expenditure that do not only affect public expenditure for organic agriculture, but also organic farming supporting schemes in general. Health care costs may also be reduced, through an increase in food quality—from the consumers' side—and a reduction in professional diseases—from the farmers' side. Furthermore, a general reduction in environmental costs might be expected from a widespread adoption of organic farming.

Output Variation

Output variation can mainly be considered under two perspectives: total—or aggregated—output variation, and crop-specific output variation.

In the first case, the most sensible aggregation rule is to convert the physical output of each crop into monetary units, in order to obtain comparable variables. Of course, in this case a bias factor is introduced in the analysis, i.e. prices; for each crop, prices can differ among countries, making comparisons and aggregation not completely reliable. When single crops are considered, then the price-bias problem is virtually avoided. Aggregation can be performed anyway in physical units within each general product category, to overcome the lack of detailed information.

Single crop output is measured multiplying the respective areas and yields. As the main objective here is to assess what would be the variation in output if the conversion to organic farming takes place according to different rates of adoption, two main sources of output variation must be considered: the difference in yields and the different land-use allocation among crops for organic and conventional farming.

The first factor should (theoretically) be quite easy to determine, by measuring the yield differences between the organic and conventional cases for each crop. Of course, heavy simplifications must be used, as a single-value yield for each crop-country is required here, hence excluding consideration of yield variations among different regions or different organic farms (including those just converted). Furthermore, another general caveat must be considered: a time-series-yield average would of course be recommended, in order to "clean" the data of statistical noise (due, for example, to weather variations), but this information is often not available.

The second factor of the analysis, i.e. the different land-use structure of organic and conventional farms, can be determined extrapolating data from existing literature. If .this is not available or is considered not sufficient, information on land use pattern for organic farming should be done by calculating the relative share of each crop in the organic and conventional regimes over the total utilisable agricultural area.

Once data on land and yields are available for the organic and total cases, output variations can be computed both by crop and eventually by country.

For each crop, the expected output is determined by:

$OrgUAA \times share_i \times OrgY_i$

where: OrgUAA = total UAA (ha) x % of hypothesized organic conversion;

share = organic area share (%) of the i-th crop;

$OrgY_i$ = organic yield of the i-th crop.

The final output variation is computed with reference to the actual total production for the i-th crop. Some problems arise within this general approach, mainly for the country-level analyses. The first one is which crops should be considered, or, in other words, which level of crop aggregation should be worked with. A further problem that also affects the analysis at the commodity level is to take into consideration if it happens that organic farms are more likely to be located in marginal areas, due e.g. to lower opportunity costs deriving from the lower yields of these areas.

This would cause problems for generalising both yields and area computations and for this reason it should be taken into consideration if the observed organic yields should be increased by a percentage that reflects the influence deriving from an unfavourable location in order to be actually comparable with the conventional yields. At the same time, an assessment of whether the (organic) marginal land would have been continued to be farmed by conventional practices is also necessary. These evaluations could be done on the basis of expert assessments and technical and agronomical considerations, focussing on the country's specific agronomical conditions.

Crop Productions

Concerning crops, when considering the direct effects that organic conversion might have on output, the attention is mainly focused on the assessment of yields reduction, as well as the different land-use patterns.

Regarding yields, there is generally no unanimous agreement on the exact loss in yield for the various crops or livestock products. Data differ quite a lot according to the specific environmental conditions, the farmers' skills, the period of conversion, the country, and so on. Furthermore, a significant lack of data is usually encountered in this field, and a quantitative indication of the yield differences between the organic and conventional cases is often missing for many crops in many countries.

The total average yield for each crop can be simply defined as: total production for i-th cro / total area of i-th crop. It can therefore be interpreted as a weighted average of the different farming types for i-th crop in a given country, where of course the prevailing farming type has the highest weight. Data on country's average yield should be available from national statistics or from FAO database. For what concerns the organic yield data, they should be taken from existing database or data of certification bodies where available, or extrapolated by existing literature. If no official or published information is available, expert assessment should be used to produce estimates of organic yields.

The reference average yield canb be of help in guiding experts to assess variation of organic yields with respect to the official yield data rather than produce a direct estimate. For example if for a certain country the average national yield of cereals is 3 tons per ha, experts may estimate that organic yield are on average 20% lower, i.e. deriving and estimate of 2,4 tons per ha.

An indication of the expected organic yield variation around the stated data may be of help for producing sensitivity analysis of total output impact using extreme yield values. These may be collected considering existing literature evidence when available, e.g. indicating the higher and lower figures found in literature for a specific crop yield, or using expert assessments.

Given the wide range of environmental conditions that may be encountered in many countries, it would be advisable to stratify yield data according to macro areas that share common basic environmental characteristics, like for instance mountain, hill and plain areas, or dry, semi-humid, humid areas, etc. For each macro-area and crop the reference yield should ideally be the average of yields observed in a sufficiently wide time frame, in order to average out biasing effects due to climatic variation and other exogenous factors.

The other main factor to consider for output variation, i.e. land use, is also difficult to evaluate. A wide range of elements need to be taken into account that can directly or indirectly affect land-use pattern changes. The most important factor is probably the specific rotation schemes adopted by organic farming, which cause extensive differences in area harvested under different crops. Furthermore, in organic farming manuring is often heavily based on livestock production, hence also causing a tendency to reallocate farm activities in order to balance livestock and crop production according to the proper organic management of the farm. This results in a different land use for the "average" organic farm, where for instance fodder crops, legumes and pasture are more prevalent than in the conventional situation.

Concerning cereals, generalisations are difficult, as some specific cereals, like oats, which can be inserted into the organic farming rotation schemes, might be cropped more widely, hence compensating for the reduction of cereal areas due to wheat and maize. Furthermore, specific products, like emmer for example, sometimes experience a renewed importance, not only with regard to rotation requirements, but also because they may help to differentiate the production, and hence to exploit the potentials of the market niches, which may be an option for exports.

Actually, market driven forces might have some indirect effect on land use for organic farming, not only through the demand for specific products, but also via different price patterns for organic products, and modifications in consumer tastes. In fact, organic price premiums, where existing, may be not evenly distributed over all products, and, of course, this can stimulate the production (i.e. land use) of those products for which premiums are higher.

Again, it is very difficult to assess to what extent and for which product in particular this factor can interact with organic land use, because this would require in-depth studies on price elasticities for the various crops, which is evidently a hard task by itself, and becomes nearly impossible given the general lack of data about organic product prices. If statistical data on organic areas are available, the "representative" organic land use pattern may be estimated considering the share of the aggregated surfaces dedicated to different productions, possibly taking into account a time frame wide enough to originate consistent averages.

A deep time length allows to average out the effects of rotation that may produce for a single year bias in the actual land use: e.g in a four course organic rotation for cereals based on two years of cereals, one year of legumes, and one year of green manure, if data are collected in the year dedicated to cereals, this would overestimate cereals weight.

It is to be taken into consideration that environmental and agronomic condition may vary considerably within a country's territory, and in such cases a generalisation at a national level of land use pattern built up on national aggregated data may represent an oversimplification of the actual situation, leading to biased results in the simulation of output variation. A stratification of land use data would be advisable, following the same approach used for yields.

If statistical data on organic areas are not available a feasible approach to the problem of estimating the different land use pattern under organic farming is to individuate for each country' macro-area the most likely rotation schemes that may be assumed as typical for the representative organic farm, and to compare it to mainstream rotation schemes of conventional farming. Information on this issues shall be obtained through existing literature or expert assessments.

An average conventional land use pattern may be computed based on national or FAO databases, and it can be modified on the basis of the information raised on the different rotation schemes. For instance, if standard conventional rotation for cereals production is based on a four course rotation with three years of cereals and one year of break crop, while the organic rotation scheme for cereals is as in the previous example two years of cereals and one year legumes and one year green manure, organic land use for cereals is reduced by 50% with respect to the conventional case.

Results of Output Variation

Once data for yield and land use are available, potential

impact of different levels of conversion to organic may be computed.

As a result, a summary scheme like Table 1 will be produced, based on Midmore and Lampkin:

Table 1 Scheme for expected output variation according to different levels of organic take up (values in,000 Tonnes)

	Overall organic take up:			
	5%	10%	...	100%
Crop 1				
Crop 2				
:				
:				
Crop n				

The basic info necessary to fill in table 1 are the organic yields for the n crops considered, and the utilisable agricultural area breakdown by crop of the country.

Two ways for proceed can be defined:

— The simple one hinges upon a strong ceteris-paribus assumption about the land use pattern of the national average and organic cases: i.e. assumes that the observed national land use pattern will not change under the different % of conversion to organic farming. Therefore, the overall output variation is simply computed using the specific organic yields with the national land use data.

— A more complex, but maybe more realistic, approach is to take into account the specific land use pattern of organic farming. In this case the organic yield data shall be used together with the expected land use pattern under the different % of organic adoption.

The overall input variation due to organic conversion may be obtained summing up the output variation computed for each macro-area. Two approaches may be followed: the first one assumes that organic conversion takes place in the same proportion in each of the country's macro areas; the second one may assume that different conversion rates may occur in different macro areas, according to more or less favourable environmental conditions for organic farming. This would allow also to model the effects of policies supporting organic conversion specifically in particular areas, like for instance less favoured areas.

The results on output implications due to organic conversion may be commented according to the country's policy for agricultural commodities. Issues to be considered are in particular:

— Development objectives

— self sufficiency for basic commodities (e.g. cereals, milk);

— foreign trade balance for agricultural commodities

Both qualitative and quantitative indicators may be developed for assessing the consequences on food self-sufficiency due to organic conversion, indicating the main areas of problems that may be expected to emerge in terms of possible import increase, ability to pay for possible income increases, food categories that may be expected to become scarce.

Macro Effects of Organic Conversion

Besides direct effects on commodities production, the conversion to organic farming may have a wide range of indirect effects on the agro-food system, on the industry of agricultural inputs and machinery, and on rural development in general. From this point of view in particular, multiple roles of organic farming should be considered.

> "Beyond its primary function of producing food and fibre, agricultural activity can also shape the landscape, provide environmental benefits such as land conservation, the sustainable management of renewable natural resources and the preservation of biodiversity, and contribute to the socioeconomic viability of many rural areas." (OECD Declaration of Agricultural Ministers Committee—1998)

There is scope for accounting for the positive "goods" that agriculture can produce beyond commodities that farmers sell in the marketplace. These goods can be defined quite broadly, but generally include rural community values such as a large number of independent, family farms, strong local economies that both rely on the economic output local farms and supply them with agricultural goods and services, rural employment, and the continued health of rural culture.

Environmental goods usually mentioned include contributions to biological diversity, clean water and air, bioenergy, and improved soils. Other products include regional or national food security, landscape values, food quality/food safety, and improvements in farm animal welfare.

Even narrowed, the analysis of indirect effects of conversion to organic takes into account a wide variety of aspects, due to linkages with the entire agro-food sector and with non agricultural sectors. A structured qualitative approach seems therefore the most effective way to take into consideration the consequences on a heterogeneous range of sectors. Multi-Sectoral Qualitative Analysis developed by Roberts and Stimson may be a basis flexible enough for such purposes, and it is already been used to assess the effects of organic farming in the agro-food market by Beckie et al.

Multi Sectoral Qualitative Analysis

Multi Sectoral Qualitative Analysis (MSQA) may be

considered as an evolution of standard SWOT analysis widely used in marketing for strategic analysis, of which maintains the qualitative approach based on expert assessments and the general aim of investigate present and potential effects, both positive and negative, of some strategic action.

The method follows an approach based on the analysis of interactions between a set of economic, social and environmental sectors and a set of territorial competences/characteristics. This kind of analysis may be of help for in depth analysis of a complex system key factors, through an investigation of relationships and linkages that play a major role.

The relevance of such interactions is measured by a qualitative judgment made by a panel of experts, and then converted into a numerical scale. As a result a set of indexes is produced, that summarise the overall role played by sectors and competences in the analysed regional context.

The analysis of interactions among sectors and territorial competences can be integrated by an evaluation of the potential risks and opportunities for the regional context under analysis, based again on qualitative expert assessments converted into numerical values and subsequently summarised in a set of index. For the purposes the original approach of Roberts and Stimson must be adapted to suit specifically the analysis of organic conversion in a specific country, and may be re-defined as Multi Stage Qualitative Analysis (MSQA). A two level approach may be adopted:

— Evaluation of competences and of the organic system specific components (supply chains of crop and livestock productions);

— Evaluation of opportunities and risks of growth of each supply chain and of the organic system as a whole.

The organic sector used in Roberts and Stimson may be referred to the more specific MSQA context of the supply chain system linked to organic production: the crop and livestock sectors must be treated separately, and for each the main elements of the supply chain can be considered:

— Main production types seed, mechanical and other input industry;

— processing industry,

— certification and labelling system,

— trade and marketing sector

Of course these are general categories, that may be further detailed according to each country specific conditions.

Competences are horizontal characteristics of the regional/national socioeconomic system that transversally affect all the sectors of the organic system. They refer therefore to the potential resources available to organise efficiently the different resources necessary for the organic sector as a whole. A possible list of competences for the organic sector may be as follows:

— economic performance;

— marketing capacity;

— technological level;

— human resources;

— management skills;

— financial issues;

— administrative issues;

— infrastructure availability

Again, competences must be defined according to the actual relevant aspects that may be encountered in the analysed country by the expert panel. Each of the competencies characterise to different extent each supply chain of the organic sector, and a joint evaluation of such interactions is the basis of MSQA analysis.

Interaction are marked in a qualitative way on the basis of a scale of attributes ranging from "weak", "medium", "strong". A higher number of qualitative values may be defined by the user, but in general no more than 5 are used. The use of qualitative judgements for assessing how a sector is "compatible" with a competence is preferred as it is usually more user friendly with respect to numerical values.

Qualitative values are then converted into numeric values. The numerical conversion allows for the computation of two indexes: competency index and sector index. The Sector Index (SI) and Competency Index (CI) are defined as:

$$SI_i = \frac{\sum_{j}^{k} S_{ji}}{k \times \max S}$$

$$CI_j = \frac{\sum_{i}^{n} S_{ji}}{n \times \max S}$$

where S_{ji} is the numerical score of the assessment of competency $_j$ with respect to sector $_i$, k is the number of competencies, n is the number of supply chains, and maxS is the maximum value that the numerical score may assume.

The indexes range between 0 and 1: the closer CI_j is to 1, the highest is the overall importance of jth competency in the country; the closer SI_i is to 1, the highest is the performance of the i-th sector measured considering all the competencies. The joint evaluation of both indexes series gives information about the filiere with the higher performances, and the competences that may be considered as crucial in the investigated system. A graphical representation may help in summarise the results for the two indexes sets.

Evaluation of potential risks and opportunities for competences and supply chain, this phase of the analysis considers how potential opportunities and risks may affect exogenously the organic system, and is therefore crucial in order to evaluate in advance possible actions that may be taken in order to avoid potential crisis or exploit at best eventual positive scenarios that may occur.

Risk and opportunity analysis is structured on a similar way as for the supply chain-competence matrix. The main difference is that supply chain are now analysed with respect to a set of potential risks and opportunities instead than a set of territorial competences. A set of g risks must be defined by the panel of experts, according to the specific situation of the country investigated.

As an example, a list of potential exogenous factors that may represent risks or opportunities for the organic system may be as follows:

— Globalisation

— economic conjuncture;

— price stability at international level

— GMOs contamination

— Technological development and innovation in the sector;

— Public opinion;

— Capital and financial resources availability;

— Skilled labour force availability;

— Natural resources availability;

— Technical assistance,

— Political climate,

— Administration stability;

Two matrixes can be generated: one refers to the evaluation of risks/opportunities for the set of n organic supply chains, the other one to the evaluation of risks/opportunities for the set of k competences.

Again, qualitative assessments for the l-th risk/opportunity and the i-th supply chain/j-th competence may be used and converted into numerical values: negative values for risk assessments, positive values for opportunities assessments. The risk/opportunity index for the i-th supply chain is defined as follows:

$$ROFI_i = \frac{\sum_{1}^{g} S_{li}}{g \times \max S}$$

ranging between -1 and 1and allows to individuate the supply chain that are expected to have more opportunities of development (ROFI close to 1) and those that are expected to face more risks (ROFI close to -1).

Similarly, a Risk/opportunity index may be computed for the l-th exogenous factor as follows:

$$ROI_i = \frac{\sum_{i}^{n} S_{ik}}{n \times \max S}$$

ranging between -1 an 1, allows to classify the exogenous factors as risks (negative ROI values) or opportunity (positive ROI values).

The very crucial aspect of MSQA is in the expert assessment data source, and therefore the selection and training of expert that will participate to the analysis must be clearly defined. Experts potentially involved must cover all the basic themes linked with the economic impact of organic farming, and should be people directly involved in the organic sector.

Micro Effects of Organic Agriculture

The aspects that may determine a variation in the profitability of a farm after conversion to organic both in positive and negative terms can be schematised in the following aspects:

— Potential yield loss;

— Change in land use due to rotational requirements;

— Price variation for organic products;

— Managerial difficulties;

— Cost variation for purchased input.

The information to be collected refers mainly to:

— structural data

— technical data

— economic data.

Individuation of Representative Organic Farm

Note that here the term "representative" is not used in strict statistical terms, representative databases of organic farms are very unlikely to be available. Therefore a "subjective" approach where on the basis of expert knowledge the farm types considered more diffuse and representative of the organic sector in the country are taken into account. For practical reason to take into account basically the following criteria; according to specific country situation such criteria may be changed or integrated.

Basic checklist for farm selection

— Dominant farm type

— Farm size mode

— Prevalent farm locations

As an example, if in a country the main organic production are arable crops and livestock, these two farm types should be taken into consideration. For each farm type (one for the livestock, one for the arable farm) the typical farm location and farm sizes should be individuated. In this simple example individuate two representative organic farms, one for each of the dominant farm types selected:

— Farm A: livestock farm, located in mountain areas with 15-20 Livestock units

— Farm B: arable farm, located in plain areas, with 15-20 ha of Utilisable Agricultural Area

Once the representative farm(s) has been individuated, it is necessary to collect the budget data required for economic analysis. A first possibility is to select a representative "average" farm trying to represent different production types and locations. Representative farms may be derived from statistical data if they are available, stratifying the total sample according to geographical area, livestock or crop production, etc. Such approach hinges upon the availability of a well structured database for organic farms, containing structural and economic information.

A second choice is to build up a set of organic farms through direct data collection, in order to make an in depth case study analysis based on real data from a limited farms sample. The selection of farm may be done taking into account different farm types and farms locations that may be considered relevant for the country investigated.

Finally, a third option is to define "virtual" representative organic farms using desk analysis and expert assessments via a guided procedure aiming to describe a consistent budget for an imaginary typical organic farm.

Data Evaluation

Once the data for the representative organic farm are available, they may be used to generate standard farm budget results, aiming to investigate gross farm income, net farm income and final profit, as indicated in Table 2. Both revenues and costs categories need to be described in detail in order to allow for in depth investigation of main factors influencing the final economic result.

The proposed scheme is based on economic budget definition, and in therefore a partial description of the farm performance given that the financial budget is missing. For the purposes and given the expected general difficulties in data collection, the economic budget approach may be considered a sufficient information base.

Table 2: Scheme for the computation of the main economic budget results

Revenues	Costs

1. Gross output
2. Variable costs
3. Gross Income = 1- 2
4. Fixed costs
5. Family Farm Income = 3- 4
6. Family wages
7. Profit = 5- 6

The general approach is quite simple: the basic elements of output and variable and fixed costs are disaggregated in terms of quantities and unit prices, and total values accordingly computed. Such scheme is the basis for different in depths analysis:

— distinction between technical and price based factors for the achievement of the main budget results;

— sensitivity analysis in terms of price variation of both inputs and output in order to individuate the critical values that assure the achievement of positive profits.

— partial budgeting analysis aiming to analyse the performance and contribution to the overall economic results of single farm productions (e.g livestock and crop productions).

— role played by subsidies using simulations concerning different combinations of price premiums and subsidies level computed subject to the achievement of positive final economic results.

Comparison of Economic Results

The issue of comparability between organic and conventional farms: The evaluation of economic performances of the different farm types considered, a further in depth analysis would be the comparison of economic results between representative organic and conventional farms. This requires the individuation of a feasible approach for the individuation and selection of comparable farm sets for the organic and conventional categories. The issue of comparability between farm system is highly problematic, as discussed among other by Lampkin.

The choice of a conventional reference farm depends on the availability of data and resources. As the objective is to isolate the effect of the farming system on profits, the choice of characteristics for the selection of comparable conventional farms has to be restricted to 'non-system determined' factors. Examples of factors that are clearly 'non-system determined' are locational factors such as region, soil texture, topography, climate and market distance.

A very simple first approach for assessing the potential income of the conventionally managed farm is therefore to use average data of a sample of conventional farms in the same region Besides theoretical issues, the problem of comparability is often conditioned by data availability. Usually a much wider sample is available for conventional than for organic farms, the latter being often described through ad-hoc direct sampling. Zanoli and Gambelli have developed an approach for such situation, based on k-means clustering. Each organic farm is used as a centroid, and comparable conventional farms are clustered around each organic farm according to the "closeness" of their characteristics with respect to the organic "centroid".

As a result, k groups of conventional farms are individuated, and for each group an "average" conventional farm is computed and can be considered as directly comparable with the respective organic centroid farm. Note that this approach can be compatible with the selection procedure of the representative organic farms: once these have been defined, if a database of conventional farms with the necessary budget and structural information is available, then conventional comparable farms may be selected.

If an adequate database for conventional farms is not available then the direct sampling or desk analysis procedures for the selection of representative organic farms may be followed, maintaining the basic characteristics of the organic representative farms in terms of farm type, size, and location.

Partial budgeting comparison between organic and conventional farms: Once a set of comparable organic and conventional farm is available, direct comparison between the two categories may be performed on the basis of the economic budget schemes above indicated.

When comparing organic and conventional farms particular attention shall be paid to the individuation of areas of relative profitability of the organic scheme. Standard quantitative results must be taken into consideration, like Gross output, Gross income, Family net Income and Profit. But besides the standard microeconomic analysis, an effort must be made in order to individuate if the organic practice may produce side positive effects for the farmer. In particular, further aspects to take into consideration are:

— number of crop and livestock products available in organic and conventional farms: this helps in assess potential risk reduction due to production diversification in organic farming; also to be taken into consideration is the eventual export potential of organic vs conventional products;

— farm gate prices for organic and conventional products: help to individuate the main area of market competitiveness of organic farming. May be considered in relation to the above mentioned point.

— evaluation of farmers family self consumption: the wider range of products may constitute a strong advantage for families heavily hinging on self produced commodities in Countries where food security is an issue. An analysis of the nature of commodities produced may help in considering nutritional benefits.

— measurement of the ratio between working capital and fixed assets: helps in assessing if organic farms may be considered less risky due to a higher share of labour and other short term input with respect to conventional farming.

Barriers to Conversion

In order to investigate factors or patterns that lead or constrain conversion decisions and to determine truthfulness and importance of those factors, qualitative methods of rapid, low cost data collection could be employed. Experience with data collection and analysis in developing countries suggests that many widely used rigorous data collection methods, particularly censuses, sample surveys, and similar quantitative methods, are not always the most appropriate for generating information impact assessment. Such methods require considerable investment of time and resources and tend to generate data that are too elaborate for their intended purposes. As a result, there has been a growing interest in the use of less rigorous methods that can provide timely information cost-effectively.

Among these methods of "rapid appraisal" three are particularly suited for analysing motivations and barriers. These are:

— focus group interviews

— in-depth cognitive interviews (laddering)

— interviews with key informants

In focus groups: participants discuss ideas, issues, insights, and experiences among themselves. Each member is free to comment, criticise, or elaborate on the views expressed by previous speakers. The role of the moderator is simply to stimulate discussion and keep it focused.

The group should be limited to 8 to 12 carefully selected participants. As far as possible, the group should be homogeneous in composition, with members sharing similar background and experience. A session generally lasts 1 to 2 hours, although in some cases it can exceed this limit.

The moderator introduces the subject, keeps the discussion going using subtle probing techniques, and tries to prevent a few participants from dominating the discussions. Specific subject areas in which focus groups have been used include: market research (where it was first used) to assess the impact or potential impact of a new product or service ; public relations; advertising ; healthcare and family planning projects; political campaigns; club member services; training evaluation ; graduate programme assessment; and research efforts.

Focus groups are well suited to discuss motivations and barriers to conversion as well as to explore potential recommendations and suggestion to overcome the latter. Other issues to consider in the discussions are: availability of information on marketing, prices, etc, such as possible increase in export opportunities and economic risk reduction due to lower investments in fixed farm costs (information systems), technology availability, cost of production, health awareness, export opportunities, certification procedures, and availability of organic pest control products and organic fertilisers.

In-depth interviews with farmers are suitable to explore farmers' behaviour and elicit their attitudes and beliefs with regards to farming activity. Using a sequential interviewing method known as "laddering" motivations that lie behind framers' choices as well as barriers to convert to organic farming can be analysed in depth. In-depth laddering interviews can be used to explore farmers' goal structures, which are strictly related to motivations since a goal is the aim or end of an action, and serves to direct behaviour.

Goals serve two motivational functions. First, they influence the direction of behaviour by expressing what to achieve and how and why the goal in question is meant to be attained. Secondly, they influence the intensity of behaviour, by determining the hierarchy of the actions to be undertaken and by expressing the weight of the criteria which represent the goals in decision-making.

The cognitive structure of farmers' decisions can be seen as a hierarchy or "means-end chain": perceived attributes of organic agriculture lead to consequences for farmers which contribute to those values which are important to them. Achieving positive consequences, which are embedded in the personal value system, can be seen as the motivating driver of a concrete behaviour; for example, the decision to convert or not.

Laddeering interviews are conducted using a special interview guide and can be also conducted using the paper-and-pencil format (hard laddering questionnaire). Using a specific software (Mecanalyst), it is possible to aggregate of the Means-End Chains (MEC) into a Hierarchical Value Map, linking different levels of goals or attributes to consequences and values.

These maps are the graphical representations of farmers' motivational structures. Key informant interviews are the most widely used rapid, low-cost data collection method.

If informants are carefully selected and appropriate interview procedures are followed, such interviews can be a source of rich, insightful data that might not be available from other sources. Simply stated, key informant interviews involve interviewing a select group of individuals who are likely to provide the needed information, ideas, and insights on a particular subject. Key informants should be carefully selected to reflect diverse viewpoints and concerns. They should be recruited from various occupational groups, socioeconomic strata, and organisations. In order to explore positive and negative factors influencing conversion to organic farming, key informants could be selected among the following groups:

— farmers, both conventional and organic;

— farmers organisations representatives;

— extension agents and advisors;

— policy-makers;

— scholars and researchers.

Key informant interviews are conducted using an interview guide that lists the topics and issues to be covered during a session. The interviewer frames the actual questions in the course of discussions. The atmosphere of these interviews is informal, resembling conversation among acquaintances.

The interviewer subtly probes the informant to elicit more information and ideas. The interviewer takes detailed notes. If all the relevant items are not covered in an interview, the researcher goes back to the informant again. Key informant interview are particularly appropriate when an understanding is required of the motivations and attitudes that direct farmers' behaviour.

Social Impact

Social impact has to do with changes in:

— Human population;

— Labour requirements & employment;

— Gender issues such as change in gender division of labour;

— Rural devleopment, land abandonment and urbanisation processes;

— Local community change;

— Human nutrition, health & public safety issues.

By social impacts "the consequences to human population of any public or private actions that alter the ways in which people live, work, play, relate to one another, organise to meet their needs and generally cope as members of society. The term also includes cultural impacts involving changes to the norms, values, and beliefs that guide and rationalise their cognition of themselves and their society.".

The International Association for Impact Assessment (IAIA) defines Social Impact Assessment (SIA) as that procedure that "includes the processes of analysing, monitoring and managing the intended and unintended social consequences, both positive and negative, of planned interventions (policies, programmes, plans, projects) and any social change processes invoked by those interventions. Its primary purpose is to bring about a more sustainable and equitable biophysical and human environment".

According to this definition:

— SIA contributes to the process of adaptive management of policies, programmes, plans and projects, and therefore needs to inform the design and operation of the planned intervention.

— SIA builds on local knowledge and utilises participatory processes to analyse the concerns of interested and affected parties. It involves stakeholders in the assessment of social impacts, the analysis of alternatives, and monitoring of the planned intervention.

The good practice of SIA accepts that social, economic and environemntal impacts are inherently and inextricably interconnected. Change in any of these domains will lead to changes in the other domains. SIA must, therefore, develop an understanding of the impact pathways that are created when change in one domain triggers impacts across other domains, as well as the iterative or flowon consequences within each domain.

A convenient way of conceptualising social impacts is as changes to one or more of the following :

— people's way of life - that is, how they live, work, play and interact with one another on a day-to-day basis;

— their culture - that is, their shared beliefs, customs, values and language or dialect;

— their community - its cohesion, stability, character, services and facilities - and their political systems - the extent to which people are able to participate in decisions that affect their lives, the level of democratisation that is taking place, and the resources provided for this purpose;

— their environment - the quality of the air and water people use; the availability and quality of the food they eat; the level of hazard or risk, dust and noise they are exposed to; the adequacy of sanitation, their physical safety, and their access to and control over resources -and their health and wellbeing—health is a state of complete physical, mental, social and spiritual wellbeing and not merely the absence of disease or infirmity;

— their personal and property rights - particularly whether people are economically affected, or experience personal disadvantage which may include a violation of their civil liberties;

— their fears and aspirations - their perceptions about their safety, their fears about the future of their community, and their aspirations for their future and the future of their children.

For the purpose of assessing the social impact of organic farming in developing countries: main indicators described below

Domain	Indicator
People's way of life	— Number of new jobs (full-time equivalents: FTE) created by organic farming: -temporary (incl. seasonal) -permanents
	— Number of new farms and enterprises started directly as organic
	— Organic farmer's household size
	— Number of female farm workers over total number of workers in organic farms (incl. farmer)
	— Number of female organic farmers over total number of organic farmers
	— Number and share of household purchasing organic food
	— Share of organic consumption over total domestic food consumption

	— Number of shops selling organic products — Number of traders (import/export) of organic products — Organic exports (in dollars or local currency) — Total organic turnover (in dollars or local currency)
Culture	— Ratio of organic farmer's household spending on culture to spending on leisure — Number of book purchased — Number of periodicals purchased — Number of children attending higher education establishments over total number of children — Number of alphabetised family members over total number of members
Community & Political system	— Number of active voluntary associations — Number of farmers cooperatives — Political participation (number of actual voters over total eligible voters) — Number of divorces over total number of marriages in the community — Number of abandoned children over total number of children born — Number of schools over total population — Numbers of doctors over total population — Number of people who require involuntary resettlement — Number of indigenous people targeted/involved — Area of indigenous settlement/total land area affected

Environment, Health & Wellbeing	— Average daily intake of kcal per household member
	— Average daily intake of proteins per household member
	— Average daily intake of vitamins per household member
	— Rate of suicides in the population
	— Rate of reported rapes and sexual attacks in the female population
	— Rate of reported family abuse over children in the children population
	— Number of working children < 14 years old over total number of children < 14 years old
	— Number of labour accidents among organic farmers
	— Prevailing diseases among organic farmers
	— Average life expectation at birth
Personal & property rights	— Size of organic farms in hectares
	— Number of organic livestock per organic farm
	— Share of owned land over total cultivated land by organic farmers
	— Degree of farmer self-sufficiency in household food production
	— Debt (in US $)/hectares in organic farms
	— Number of bankrupt organic farmers over total number of organic farmers
Fears & aspirations	— Main reported safety hazards among organic farmers
	— Main reported fears among organic farmers
	— Main reported aspirations of organic farmers

Some of these indicators could be used for comparisons between organic rural "districts" (where there is a significant proportion of organic farms and enterprises) and conventional districts.

Critical Social Indicators

The appropriate social indicators on the basis of availability of quantitative measurements. These could be chosen among those presented in the previous table or supplemented by others agreed by the Assessment Team.

The first step involves identifying those social impact domains that represent critical items relative to the human environment. Development of organic farming is often expected to have neutral impact on many domains since it stems from previous conventional agricultural practices, but in some marginal areas it may have a positive impact on population and therefore a negative impact on the public services. To identify these critical impact areas generally requires:

a. the selection of appropriate indicators on the basis of availability of data. A participatory approach, involving farmers and other stakeholders, is particularly recommended.

b. a comparison of expected future service levels with existing service levels, as well as the use of local planning factors and community design standards to assess the needs of the new development and its occupants.

Data Collection

Collect pertinent data and information that will enable description of the context in terms of various selected social impact domains such as those described above. Accomplishing this step will generally require the use of various sources of information such as census data, statistical abstracts of various governmental agencies, planning agencies, chambers of commerce, university research departments & libraries, and others.

Detailed data collection and analysis are needed to clarify the expected impact and effects of projects or strategies for the development of organic farming on different groups, including the poor and excluded.

Changes in Social Indicators

This requires that one have various tools, techniques, and methodologies for making these predictions. Once a number of relevant impact indicators are selected for each social domain, the third step involves the quantitative prediction, or at least qualitative description, of the changes in the social environment as the result of the new organic development. The measurement and description of the current situation, of the deadweight and of the potential impact foreseen with different level of conversion to organic farming on the society should be achieved mainly by means of desk research, supplemented—if the budget allows it—by other qualitative and quantitative methods of data collection.

In order to estimate the probable social impact of organic farming in the future, futurefocus methods could be used such as:

— *Straight-line trend impact analysis:* projecting current trends, such as population change or employment, into the future (with or without modifying the rate of change). Its simplest form is a quantitative statistical forecasting model enriched by qualitative assessments, that allows the definition of possible events that might modify the estimated trends. This approach turns out to be particularly effective for at least two reasons: it combines traditional and qualitative forecasting techniques, and stimulates analysts and experts to take into account possible effects of "unusual" events. Nevertheless, it does not take into account effects of interrelationships among variables. A further limitation of this method is the fact that it emphasizes the predetermined trends and allows only limited account for uncertain events. But it is very simple to use and can be easily combined with any of the following.

— *Comparative case study:* examining how an affected community has responded to change in the past, or the impact on other communities that have undergone a similar action. Traditional approaches to rural development focus primarily on economic issues, such as the improvement of output, income and employment. When connected to the concept of sustainability, however, a rather more multifaceted perspective has emerged, not only in relation to the conservation of environmental resources, but also social and cultural dimensions. The case study approach allows mapping of complex interrelationships in rural development to describe functional transformations in the use of resources such as land, labour, knowledge and underlying nature. Using the framework of Actor Network Theory (ANT), case study can expose the asymmetry of power relations in a rural network.

Actors are entities able to connect material, social or cultural artefacts to create a life-world filled with other entities, having their own history, identity and relations. Between them are intermediaries, the means by which actors are defined in the interaction, and the Actor Network consists of actors and intermediaries which transmit relationships. Using this framework, comparative case study allows to develop a mapping framework to describe the dynamics of power relations, showing actors as nodes and mapping them across both horizontal and vertical network relations. This mapping concerns relationships between people, and allows to explore the social effects of any new rural development, such as that related to organic farming.

— *Delphi:* The Delphi approach involves successive questionnaires to an expert panel, using feedback to refine an infor"obtain the most reliable consensus of opinion of a group of experts...by a series of intensive questionnaires interspersed with controlled opinion feedback". Delphi is not a procedure intended to challenge statistical or model-based procedures, against which human judgment is generally shown to be inferior: it is intended for use in judgment and forecasting situations in which pure model-based statistical methods are not practical or possible because of the lack of appropriate quantitative data, and thus where some form of human judgmental input is necessary.

  Four key features may be regarded as necessary for defining a procedure as a 'Delphi'. These are: anonymity, iteration, controlled feedback, and the statistical aggregation of group response. Anonymity is achieved by means of questionnaires, which are iterated over the same panel without allowing undue interactions and pressures to emerge, contrary to what is obtained by the means of focus groups. Between each iterations, respondents are informed of the opinions of their anonymous colleagues, thus providing controlled feedback.

— *Scenario analysis:* scenarios are logical-imaginations based on construction of hypothetical futures through a process of mentally modelling the assumptions about the variables in question; experts can be asked to analyse different scenarios and assess their implications on social indicator. Scenarios are helpful to develop the likely, alternative or preferred future of a community or society. Scenarios can be used to compare different outcomes. Scenario analysis usually is based on following steps:

1. Develop a common framework of analysis. Define the issues that should be understood better in terms of time frame, scope and decision variables.
2. Exploratory analysis. Using STEEP (analysis, make a list of current trends and predetermined elements that will affect the variables of interest. Construction of a flow-chart of interlinkages and causal relationships. Identify uncertainties and examine interrelations among the uncertain events. Identify the critical decision-makers.
3. Plot the scenarios. Identify the key uncertainties (driving forces) likely to have relevant impact on specific social context. Assess the degree of uncertainty and the impacts of different driving forces, in order to build a set of different scenarios. Identify events that could trigger the sequence of the events, suing cause-effect analysis. Select the most probable sequences of the events and decisions.
4. Write the scenarios. The scenarios are first drafted in skeleton form, to highlight their main ingredients. Once accepted by the Assessment Team, they should be expressed and painted in detail, by a verbal narrative describing the events and showing how the future might evolve in that direction.
5. Stakeholder analysis. Identify stakeholders and key

players in each scenario. Analyse the likely impact of each scenarios on them and on their decisions.

— Assess the internal consistency and plausibility of the scenarios. Identify where and why the scenarios may be incoherent or internally inconsistent (in terms of trends and outcome combinations). Eliminate combinations that are not credible or impossible, and create new scenarios until internal consistency is achieved.

Implications of Changes

The final step in the prediction and assessment of the impacts on the social environment involves a discussion of the implications of these changes for each domain area impacted. These implications could be examined by comparing the values of the impact measures used, such as students generated prior to development versus students generated after development, or number of hospital treatments required prior to development versus number of hospital treatments required after development. Differences in these values would mean either an increase or decrease in some type of service, demographic characteristic, or other social indicator.

10

Composting Methods

Composting is the natural process of 'rotting' or decomposition of organic matter by microorganisms under controlled conditions. Raw organic materials such as crop residues, animal wastes, food garbage, some municipal wastes and suitable industrial wastes, enhance their suitability for application to the soil as a fertilising resource, after having undergone composting.

Compost is a rich source of organic matter. Soil organic matter plays an important role in sustaining soil fertility, and hence in sustainable agricultural production. In addition to being a source of plant nutrient, it improves the physico-chemical and biological properties of the soil.

As a result of these improvements, the soil: (i) becomes more resistant to stresses such as drought, diseases and toxicity; (ii) helps the crop in improved uptake of plant nutrients; and (iii) possesses an active nutrient cycling capacity because of vigorous microbial activity. These advantages manifest themselves in reduced cropping risks, higher yields and lower outlays on inorganic fertilisers for farmers.

TRADITIONAL METHODS

Anaerobic Decomposition

Indian bangalore method

This method of composting was developed at Bangalore in India by Acharya. The method is basically recommended when night soil and refuse are used for preparing the compost. The method overcomes many of the disadvantages of the Indore method such as problem of heap protection from adverse weather, nutrient losses due to high winds / strong sun rays, frequent turning requirements, fly nuisance etc. but the time involved in production of a finished compost is much longer. The method is suitable for areas with scanty rainfall.

Preparation of the pit: Trenches or pits about one metre deep are dug; the breadth and length of the trenches can be made depending on the availability of land and the type of material to be composted. The selection of site for the pits is made as in the Indore method. The trenches should preferably have sloping walls and a floor of 90-cm slope to prevent water logging.

Filling the pit: Organic residues and night soil are put in alternate layers and, after filling, the pit is covered with a 15-20 cm thick layer of refuse. The materials are allowed to remain in the pit without turning and watering for three months. During this period, the material settles down due to reduction in volume of the biomass and additional night soil and refuse are placed on top in alternate layers and plastered or covered with mud or earth to prevent loss of moisture and breeding of flies. After the initial aerobic composting which is for about eight to ten days, the material undergoes anaerobic decomposition at a very slow rate and it takes about six to eight months to obtain the finished product.

Passive composting

Passive composting involves simply stacking the materials in piles to decompose over a long time period with little agitation and management. The process has been used for composting of animal wastes. Needless to mention that simple placement of manure in a pile does not satisfy the requirements for continuous aerobic composting. Without considerable bedding material, the moisture content of manure exceeds the level, which enables an open porous structure to exist in the pile. Little, if any, air passes through it.

Under these circumstances, the anaerobic micro-organisms dominate the degradation. All of the undesirable effects associated with anaerobic degradation occur including low temperatures, slow decomposition, and the release of hydrogen sulphide and other malodorous compounds.

When a livestock management system relies on bedding to add to livestock comfort and cleanliness, the bedding becomes mixed with the manure and creates a drier, more porous mixture. This provides some structure and, depending on the amount of bedding, enables the mixture to be stacked in true piles. The bedding also tends to raise the C:N ratio of the manure.

A mixture of manure and bedding requires a considerable proportion of bedding to provide the porosity necessary for composting. At least equal volumes of bedding and manure are required. If the amount of bedding is too low to provide a porous mix, then additional dry amendments must be provided by either increasing the bedding used in the barn or adding amendments when piles are formed. Manure from horse stables or bedded manure packs can often compost in piles alone, whereas non-bedded manure from dairy, swine, and many poultry barns needs drying or additional amendments.

Aerobic Decomposition through Passive Aeration

Indian indore method

An important advance in the practice of composting was made at Indore in India by Howard during the period 1924 to 1926. The traditional procedure was systematised into a method of composting now known as the 'Indore method'.

Raw materials: The raw materials used are mixed plant residues, animal dung and urine, earth, wood ash and water. All organic material wastes available on a farm such as weeds, stalks, stems, fallen leaves, prunings, chaff, fodder leftovers and so on, are collected and stacked in a pile. Hard woody material like cotton or pigeon pea stalks and stubble are first spread on the farm road and crushed under vehicles such as tractors or bullock carts before being piled. Such hard materials should in any case not exceed ten percent of the total plant residues. Green materials, which are soft and succulent, are allowed to wilt for two to three days to remove excess moisture before stacking; they tend to pack closely if they are stacked in the fresh state.

The mixture of different kinds of organic material residues ensures a more efficient decomposition. While stacking, each type of material is spread in layers about 15 centimetres thick until the heap is about one and a half metres high. The heap is then cut into vertical slices and about 20-25 kilograms are put under the feet of cattle in the shed as bedding for the night. The next morning the bedding, along with the dung and urine and urine-earth, is taken to the pits where the composting is to be done.

Site and pit dimension: The site selected for the compost pit should be at high level so that no rainwater gets in during the monsoon season; it should be near to the cattle shed and a water source. A temporary shed may be constructed over it to protect the compost from heavy rainfall. The pit should be about 1 m deep, 1.5-2 m wide and of any suitable length.

Filling the pit: The material brought from the cattle shed is spread evenly in the pit in layers of 10-15 cm. On each layer is spread a slurry made with 4.5 kg dung, 3.5 kg urine-earth and 4.5 kg of inoculum taken from a 15 day-old composting pit. Sufficient quantity of water is sprinkled over the material in the pit to wet it. The pit is filled in this way, layer by layer, and it should not take longer than one week to fill. Care should be taken to avoid compacting the material in any way.

Turning: The material is turned three times during the whole period of composting; the first time 15 days after filling the pit, the second after another 15 days and the third after another month. At each turning, the material is mixed thoroughly, moistened with water and replaced in the pit.

Site and heap dimensions: During rainy seasons or in regions with heavy rainfall, the compost may be prepared in heaps above ground and protected by a shed. The basic Indore pile is about 2 m wide at the base, 1.5 m high and 2 m long. The sides are tapered so that the top is about 0.5 m narrower in width than the base. A small bund is sometimes built around the pile to protect it from wind, which tends to dry the heap.

Forming the heap: The heap is usually started with a 20 cm layer of carbonaceous material such as leaves, hay, straw, sawdust, wood chips and chopped corn stalks. This is then covered with 10 cm of nitrogenous material such as fresh grass, weeds or garden plant residues, fresh or dry manure or digested sewage sludge. The pattern of 20 cm carbonaceous material and 10 cm of nitrogenous material is followed until the pile is 1.5 m high and the material is normally wetted so that it may feel damp but not soggy. The pile is sometimes covered with soil or hay to retain heat and is turned at six- and twelve-week intervals. In the Republic of Korea, heaps are covered with thin plastic sheets to retain heat and prevent insect breeding. If materials are limited, the alternate layers can be added as they become available.

Chinese rural composting

The pit method: Generally, composting is carried out in a corner of a field and in a circular or rectangular pit. Rice straw, animal dung (usually pig), aquatic weeds or green manure crops are used and often silt pumped from river beds is mixed with the crop residues. The pits are filled layer by layer, each layer being 15 cm thick. Usually, the first layer is of a green manure crop or water hyacinth, the second layer is a straw mixture and the third layer is of animal dung. These layers are alternated until the pit is full, when a top layer of mud is added; a water layer of about 4 cm depth is maintained on the surface to create anaerobic conditions which help to reduce losses of nitrogen.

Approximate quantities of the different residues in tons per pit are: river silt 7.5, rice straw 0.15, animal dung 1.0, aquatic plants or green manure 0.75 and superphosphate 0.02. Three turnings are given in all, the first one month after filling the pit and, at this time, the superphosphate is added and thoroughly mixed in. Water is added as necessary. The second turning is done after another month and the third two weeks later. The material is allowed to decompose for three months and produces about eight tons of compost per pit.

High temperature compost: This form of compost is prepared mainly from night soil, urine, sewage, animal dung, and chopped plant residues at a ratio of 1:4. The materials are heaped in alternate layers starting with chopped plant stalks and followed by human and animal wastes; water is added to optimum amount.

At the time of making the heap, a number of bamboo poles are inserted for aeration purposes. After the heap formation is complete, it is sealed with 3 cm of mud plaster. The bamboo poles are withdrawn on the second day of composting leaving the holes for aeration of the heap. Within four to five days, the temperature rises to 60-70C and the holes are then sealed. The first turning is usually done after two weeks and the moisture is made up with water or animal or human excreta; the turned heap is again sealed with mud. The compost is ready for use within two months.

Large Scale Passive Aeration

Windrow Composting

Turned Windrows: Windrow composting consists of placing the mixture of raw materials in long narrow piles or windrows which are agitated or turned on a regular basis. The turning operation mixes the composting materials and enhances passive aeration. Typically the windrows are initially from 3 feet high for dense materials like manures to 12 feet high for fluffy materials like leaves. The width varies from 10 to 20 feet. The equipment used for turning determines the size, shape, and spacing of the windrows. Bucket loaders with a long reach can build high windrows. Turning machines produce low, wide windrows.

Windrows aerate primarily by natural or passive air movement. The rate of air exchange depends on the porosity of the windrow. Therefore, the size of a windrow that can be effectively aerated is determined by its porosity. A light fluffy windrow of leaves can be much larger than a wet dense windrow containing manure. If the windrow is too large, anaerobic zones occur near its centre which release odours when the windrow is turned.

For small to moderate scale operations, turning can be accomplished with a front end loader or a bucket loader on a tractor. The loader simply lifts the materials from the windrow and spills them down again, mixing the materials and reforming the mixture into a loose windrow. The loader can exchange material from the bottom of the windrow with material on the top by forming a new windrow next to the old one. This needs to be done without driving onto the windrow in order to minimize compaction. Windrows turned with a bucket loader are often constructed in closely spaced pairs and then combined after the windrows shrink in size. If additional mixing of the materials is desired, a loader can also be used in combination with a manure spreader.

A number of specialised machines have been developed for turning windrows. These machines greatly reduce the time and labour involved, mix the materials thoroughly, and produce a more uniform compost. Some of these machines are designed to attach to farm tractors or front-end loaders; others are self-propelled. A few machines also have the capability of loading trucks or wagons from the windrow.

It is very important to maintain a schedule of turning. The frequency of turning depends on the rate of decomposition, the moisture content and porosity of the materials, and the desired composting time. Because the decomposition rate is greatest at the start of the process, the frequency of turning decreases as the windrow ages. Easily degradable or highnitrogen mixes may require daily turnings at the start of the process. As the process continues, the turning frequency can be reduced to a single turning per week.

By the end of the first week of composting, the windrow height diminishes appreciably and by the end of the second week it may be as low as 2 feet. It may be prudent to combine two windrows at this stage and continue the turning schedule as before. Consolidation of windrows is a good wintertime practice to retain the heat generated during composting. This is one of the advantages of windrow composting. It is a versatile system that can be adjusted to different conditions caused by seasonal changes.

Passively aerated windrows: The method, passively aerated windrow system, eliminates the need for turning by supplying air to the composting materials through perforated pipes embedded in each windrow. The pipe ends are open. Air flows into the pipes and through the windrow because of the chimney effect created as the hot gases rise upward out of the windrow. The windrows should be 3-4 feet high, built on top of a base of straw, peat moss, or finished compost to absorb moisture and insulate the windrow. The covering layer of peat or compost also insulates the windrow; discourages flies; and helps to retain moisture, odour, and ammonia. The plastic pipe is similar to that used for septic system leach fields with two rows of ½-inch diameter holes drilled in the pipe. In many aerated pile applications, the pipe holes are oriented downward to minimize plugging and allow condensate to drain. However, some researchers recommend that the holes face upwards.

Windrows are generally formed by the procedures described for the aerated static pile method. Because the raw materials are not turned after the windrows are formed, they must be thoroughly mixed before they are placed in the windrow. Avoid compacting the mix of materials while constructing the windrow. Aeration pipes are placed on top of the peat/compost base. When the composting period is completed, the pipes are simply pulled out, and the base material is mixed with the compost.

Rapid Method

While traditional composting procedures take as long as 4-8 months to produce finished compost, rapid composting methods offer possibilities for reducing the processing period up to three weeks. A variety of approaches and their combinations have been used to hasten the composting process, which include the following:

- — Shredding and frequent turnings
- — Use of chemical nitrogen activators
- — Use of Effective Microorganisms (EM)
- — Use of worms
- — Use of celluloytic cultures
- — Use of forced aeration
- — Use of forced aeration and mechanical turnings

Frequent Turnings

Berkley rapid composting

The method corrects some of the problems associated with the old type of composting. With this process, compost can be made in two to three weeks. There are several important factors essential to the rapid composting method given below:

— Material will compost best if it is between ½ to 1 ½ inches in size. Soft, succulent tissues need not be chopped in very small pieces because they decompose rapidly. The harder or the more woody the tissues the smaller they need to be divided to decompose rapidly. Woody material should be put through a grinder, but most grinders chop herbaceous materials too finely for good composting. Chopping material with a sharp shovel is effective. When pruning plants, cut material into small pieces with the pruning shears-it takes a little effort but the results are good.

— For the composting process to work most effectively, material to be composted should have a carbon to nitrogen ratio of 30:1. Mixing equal volumes of green plant material with equal volumes of naturally dry plant material will give approximately a 30:1 carbon to nitrogen (C/N) ratio. Green material can be grass clippings, old flowers, green prunings, weeds, fresh garbage and fruit and vegetable wastes. Dried material can be dead, fallen leaves, dried grass, straw and somewhat woody materials from prunings.

— Materials, which should not be added to a composting pile, include soil, ashes from a stove or fireplace, and manure from carnivorous (meat-eating) animals. Manures from herbivorous animals such as rabbits, goats, cattle, horses, elephants or fowl can be used. Once a pile is started, do not add anything. The reason is that it takes a certain length of time for the material to break down and anything added has to start at the beginning, thus lengthening the decomposition time for the whole pile. Excess material should be as dry as possible during storage until a new pile is started. Moist stored materials will start to decompose and if this occurs, they will not do a good job in the compost pile. Nothing needs to be added to the organic materials to make them decompose. The microorganisms active in the decomposition process are ubiquitous where plant materials are found and will develop rapidly in any compost piles.

— Composting works best if the moisture content of materials in the pile is about 50 percent. Too much moisture will make a soggy mass, and decomposition will be slow and will smell. If the organic material is too dry, decomposition will be very slow or will not occur at all.

— Heat, which is very important in rapid composting, is supplied by the respiration of the microorganisms as they break down the organic materials. To prevent heat loss and to build up the amount of heat necessary, a minimum volume of material is essential: a pile at least 36" X 36" X 36" is recommended. If less than 32 inches, the rapid process will not occur. Heat retention is better in bins than in open piles, so rapid composting is more effective if bins are used. In addition, the use of bins is much neater. High temperatures favour the microorganisms which are the most rapid decomposers; these microorganisms function at about 160F (71C) and a good pile will maintain itself at about that temperature. A thermometer to measure temperatures inside the pile is helpful although not necessary.

— The compost pile needs to be turned to prevent the pile from getting too hot. If it gets much above 160F, the microorganisms will be killed, the pile will cool, and the whole process will have to start from the beginning. By turning the pile it will not overheat, and it will be aerated also, both of which are necessary to keep the most active decomposers functioning. The pile should be turned so that material which is on the outside is moved to the centre. In this way, all the material will reach optimum temperatures at various times. Due to heat loss around the margins, only the central portion of the pile is at the optimum temperature. Because of the necessity for turning, it is desirable to have two bins so the material can be turned from one into another. Bins made with removable slats in the front make the turning process easier. Bins with covers retain the heat better than do those having no covers. Once the decomposition process starts, the pile becomes smaller and because the bin is no longer full, some heat will be lost at the top. This can be prevented by using a piece of polyethylene plastic slightly larger than the top area of the bins. After the compost is turned, the plastic is placed directly on the top of the compost and is tucked in around the edges. If the material in the pile is turned every day, it will take 2 weeks or a little longer to compost. If turned every other day, it will take about 3 weeks. The longer the interval between turning the longer it will take for the composting to finish.

— If done correctly, a pile will heat to high temperatures within 24 to 48 hours. If it doesn't, the pile is too wet or too dry or there is not enough green material (or nitrogen) present. If too wet, the material should be spread out to dry. If too dry, add moisture. If neither of these, then the nitrogen is low (a high C/N ratio) and this can be corrected by adding materials high in nitrogen.

— If the C/N ratio is less than 30:1, the organic matter will decompose very rapidly but there will be a loss of nitrogen. This will be given off as ammonia and if this odour is present in or around a composting pile, it means that valuable nitrogen is being lost in the air. This can be counteracted by the addition of some sawdust to that part of the pile where there is an ammonia odour-sawdust is very high in carbon and low in nitrogen (a high C/N ratio) and therefore will counteract the excess nitrogen. Other than adding water should the pile become dry, this is the only thing, which should be added to a pile once it's started. Because composting can be done anytime, during the rainy season some covering the pile may be necessary to keep the composting materials from becoming too wet.

— The rapid decomposition can be detected by a pleasant odour, by the heat produced, by the growth of white fungi on the decomposing organic material, by a reduction of volume, and by the change in colour of the materials to dark brown. As composting nears completion the temperature drops and, finally, little or no heat is produced. The compost is then ready to use. If in the preparation of the compost the material was not chopped in small pieces, screening the material through 1-inch-mesh chicken wire will hold back such pieces. These can be added to the next pile and eventually they will decompose.

Mineral Nitrogen Activators

North Dakota State University Hot Composting

Compost piles with a height of 6 feet (1.8 m) are raised. The maximum size of the organic matter pieces should be 6-9 inches (15-23 cm) long. If bins are constructed, the dimensions of 5x5x6 feet will yield 150 cubic feet (4.3 cu.m) of compost. This is a respectable volume of compost to produce in a four- to six-week time period. To keep the aerobic bacteria population high and active, 0.12 kg of actual nitrogenous fertiliser should be added per cubic foot of dry matter and holes punched (four to five) into the centre of the pile. This is best done in phases or stages as the compost pile is building up. The total should be about 17-18 kg of fertiliser for the entire pile.

In this high-temperature, bacterially active system, it is best to turn the composting material every three to four days. Once activated, expect the temperature to range between 120-160°F (49-71°C). The decomposition will go faster in summer (as short as three to four weeks) and take more time in the spring and fall. No measurable activity occurs during typical North Dakota winters. Once the compost is no longer hot and is an odour-free, crumbling material, it is ready for use.

Effective Microorganisms (EM)

EM based compost production

Since 1999 seven small-scale organic fertiliser plants, using the EM based quick production process have been in operation in Myanmar. They are owned and operated by Women's Income Generation Groups (WIGG). A unit plant consists of nine pits of 6 ft (l) x 4 ft (w) x 3 ft (d), enclosed by low walls and covered with roof.

Raw Materials:

The raw materials for organic fertiliser production are as follows:

—	Cow dung	2 portions
—	Rice husk	1 portion
—	Rice husk/charcoal	1 portion
—	Rice bran, milled	1 portion
—	Accelerator	33 litres of EM solution or Trichoderma solution per pit.

Preparation of EM solution (accelerator): Firstly one litre of 'instant solution' is made by mixing 10 ml EM, 40 ml molasses and 950 ml water and leaving it for five to seven days, depending on temperature. Then the solution is added to one litre of molasses and 98 litres of water to obtain 100 litres of readyto-use EM solution. This amount is enough for three pits. The EM solution functioning as accelerator reduces the composting period from three months to one month.

Procedure: Firstly, mix all the ingredients, except accelerator. Then make 0.5 ft layer of mixture in the pit and sprinkle accelerator over. Repeat the same procedure until the pit is full. Cover with plastic sheet. Two or three weeks later, mix the whole pit to boost aerobic decomposition. The fertiliser is ready to use a couple of weeks later.. A pit turns out 900 kg of final product per batch, which are usually packed in 30 kg plastic bags. Assuming that it takes 30 days on average to produce a batch and only eight pits may be used for technical reasons, the annual potential production capacity works out to 86.4 tons.

Celluloytic Cultures

IBS Rapid Composting Technology

IBS Rapid Composting: IBS rapid composting technology involves inoculating the plant substrates used for composting with cultures of Trichoderma harzianum, a cellulose decomposer fungus. The fungus, grown in a medium of sawdust mixed with the leaves of ipil ipil (Leucaena leucocephala), a leguminous tree, is termed as compost fungus activator (CFA). The technology is a development of the windrow type of composting. The composting time, using this procedure, ranges from 21 to

45 days, depending on the plant substrates used. The procedure consists of two parts: the production of the compost fungus activator, and the composting process.

Preparation of substrates: Substrates such as rice straw, weeds and grasses should be chopped. Chopping helps speed up decomposition by increasing the surface area available for microbial action, and providing better aeration. If large quantities of substrates are to be used, a forage cutter/chopper is needed. Chopping can be dispensed with if the compost is not needed in the near future.

Adjustment of moisture content: Substrates should be moistened with water. Plant substrates can be soaked overnight in a pond, which cuts down on the need for water. If a large volume of substrates arc to be composted, a sprinkler is more convenient.

The compost mixture: Carbonaceous substrates should be mixed with nitrogenous ones at a ratio of 4:1 or less, but never lower than 1:1 (on a dry weight basis). Some possible combinations are:

— 3 parts rice straw - 1 part ipil-ipil

— 4 parts rice straw - 1 part chicken manure

— 4 parts grasses - 1 part legume materials + 1 part manure

— 4 parts grasses - 1 part Chromolaena odorata or Mikania cordata + 1 part animal manure

Composting procedure

— The substrates should be piled loosely in a compost pen to provide better aeration within the heap. The material should not be too compact and no heavy weights should be put on top. Compost heaps should be located in shady areas such as under big trees. The platform should be raised about 30 cm from the ground, to provide adequate aeration at the bottom. Alternatively, aeration can be provided by placing perforated bamboo trunks horizontally and vertically at regular intervals, to carry air through the compost heap.

— The compost activator, consisting of a cellulolytic fungus, is broadcast onto the substrates during piling. The amount of activator used is usually 1% of the total weight of the substrates. Decomposition is faster if the activator is mixed thoroughly with the substrate. A greater amount of activator can be used if faster decomposition is desired.

— The heap should be covered over completely. This maintains the heat of decomposition, and minimizes water evaporation and ammonia volatilisation. White plastic sheets, or plastic sacks with their seams opened and sewn together, can serve as a cover. The compost heap usually heats up in 24 - 48 hours.

— Heat should be maintained at 50°C or higher, and the heap should be turned over every 5-7 days for the first two weeks, and thereafter once every two weeks. After the first week, the volume of the pile should be reduced by one-third. After two weeks, the volume of the pile should be reduced to one half the original.

— The mature compost should be removed from the pen, and dried in the sun for two days. It should then be put into sacks and stored in a shaded area. Decomposition will continue until the substrate is finely fragmented, so that the finished product has a powdery texture. Then, once decomposition is complete, the compost should be sundried again until the moisture content is at 10-20%.

— If mature compost is needed at once, it should be sun-dried for one day, as soon as its temperature drops to 30°C. Drying removes excess moisture, and makes the compost much easier to handle. Although the compost still retain some fibres, it can be applied immediately as fertiliser.

IBS Rapid Commercial Compost Production: In the large-scale commercial production of compost, the following operations need to be mechanised, other steps remaining the same:

— Chopping of substrates.

— Mixing/Turning - when there are several tons of substrate, a pay loader will make mixing of substrates or turning of heaps much easier.

— A hammer mill should be used to break up big lumps of mature compost before drying.

— During rainy months, it is more economical to dry compost mechanically than try to sun dry it.

IBS rapid rice straw composting

At harvest time, rice straw is heaped on to one side of the paddy field. It saves labour to have one compost pile for each paddy field instead of one central pile. Various steps are as follows:

— Rice straw is soaked overnight in water or in the rain until saturated.

— A simple platform is made in the middle of the field

— A layer of saturated rice straw 10-15 cm thick is loosely piled on the platform.

— On top of the layer, one or two handfuls of the activator is broadcast (25 kg /ha).

— Straw is alternately layered with the activator until all the straw has been used.

— Manure and nitrogenous plants are put on top of the straw layers. The nitrogen substrate is 15-25% of total composition.

— The compost is covered (with plastic, banana leaves, or coconut fronds) and heats up within 25 hours.

— The compost must be moistened frequently to compensate for evaporation.

— The compost is left unturned and matures within one month. It is ready for use when the pile has cooled and is 30% of its original size.

Forced Aeration

Aerated static pile method

The aerated static pile method takes the piped aeration system a step further, using a blower to supply air to the composting materials. The blower provides direct control of the process and allows larger piles. No turning or agitation of the materials occurs once the pile is formed. When the pile has been properly formed and if the air supply is sufficient and the distribution is uniform, the active composting period will be completed in approximately three to five weeks.

With the aerated static pile technique, the raw material mixture is piled over a base of wood chips, chopped straw, or other very porous material. The porous base material contains a perforated aeration pipe. The pipe is connected to a blower, which either pulls or pushes air through the pile.

The initial height of piles should be 5-8 feet high, depending on the material porosity, weather conditions, and the reach of the equipment used to build the pile. Extra height is advantageous in the wintertime to retain heat. It may be necessary to top off the pile with 6 inches of finished compost or bulking agent. The layer of finished compost protects the surface of the pile from drying, insulates it from heat loss, discourages flies, and filters ammonia and potential odours generated within the pile.

Two forms of aerated static piles are common: individual piles and extended piles. Individual piles are long triangular piles with a width (10-16 feet, not including the cover) equal to about twice the pile height. The aeration pipe runs lengthwise beneath the ridge of the pile. Individual piles hold a single large batch of material or a few batches of roughly the same recipe and age (within three days, for example). Individual piles are practical when raw materials are available for composting at intervals rather than continuously.

Since the pile does not receive additional turnings, the selection and initial mixing of raw materials are critical. Otherwise, poor air distribution and uneven composting occur. The pile must have good structure as well as to maintain porosity through the entire composting period. This generally requires a fairly stiff bulking agent such as straw or wood chips. Wood chips are commonly used for composting sewage sludge by this method. Because of their large size, wood chips pass through the process only partially composted. They are usually screened from the finished compost and reused as bulking agents for an additional two or three cycles.

Since straw decomposes over the composting period, a pile with straw as an amendment can gradually lose structure. This is partially compensated by the drying which takes place as composting proceeds. Other possible bulking agents and amendments for static pile composting include recycled compost, peat moss, corn cobs, crop residues, bark, leaves, shellfish shells, waste paper, and shredded tires. Uncomposted material like shredded tires and mollusc shells must eventually be screened from the compost and reused. To obtain good air distribution, manure or sludge must be thoroughly blended with the bulking agent before the pile is established.

The required airflow rates and the choice of blowers and aeration pipe depend on how aeration is managed—that is, how the blower is controlled. The blower can be controlled in several different modes. It can be run continuously or intermittently. In the latter case, the control mechanism can be either a programmed time clock or a temperature sensor. The airflow rates are based on the dry weight of the primary raw material, such as sludge or manure. They should take into account the presence of typical amendments like wood chips, straw, and compost. In practice, it may be necessary to adjust the timer cycle, pile size, or blower, to suit the specific conditions and materials.

For static pile composting, the air can be supplied in two ways: a suction system with the air drawn through the pile or a pressure system with the blower pushing the air into the pile. Suction draws air into the pile from the outer surface and collects it in the aeration pipe. Since the exhaust air is contained in the discharge pipe, it can be easily filtered if odours are occurring during the composting process.

Forced Aeration and Accelerated Mechanical Turnings

In-Vessel composting method

In-vessel composting refers to a group of methods which confine the composting materials within a building, container, or vessel. In-vessel methods rely on a variety of forced aeration and mechanical turning techniques to speed up the composting process. Many methods combine techniques from the windrow and aerated pile methods in an attempt to overcome the deficiencies and exploit the attributes of each method.

There are a variety of in-vessel methods with different combinations of vessels, aeration devices, and turning mechanisms.

Bin Composting: Bin composting is perhaps the simplest in-vessel method. The materials are contained by walls and usually a roof. The bin itself may simply be wooden slatted walls (with or without a roof), a grain bin, or a bulk storage building. The buildings or bins allow higher stacking of materials and better use of floor space than freestanding piles. Bins can also eliminate weather problems, contain odours, and provide better temperature control.

Essentially, bin composting methods operate like the aerated static pile method. They include some means of forced aeration in floor of the bin and little or no turning the materials. Occasional remixing of material in the bins can invigorate the process. If several bins are used, the composting materials can be periodically moved from one bin to the next in succession. Most of the principles and guidelines suggested for the aerated pile should apply to bin composting as well. One exception relates to relatively high bins. In this case, there is a greater degree of compaction and a greater depth of materials for air to pass through. Both factors increase the material's resistance to airflow (pressure loss). A raw material with a stronger structure and/or a higher pressure blower may be required, compared to the aerated static pile method.

Rectangular Agitated Beds: The agitated bed system combines controlled aeration and periodic turning. In this system, composting takes place between walls which form long, narrow channels referred to as beds. A rail or channel on top of each wall supports and guides a compost-turning machine. Raw materials are placed at the front end of the bed by a loader. As the turning machine moves forward on the rails, it mixes the compost and discharges the compost behind itself. With each turning, the machine moves the compost a set distance toward the end of the bed. The turning machines work much like windrow turners, using rotating paddles or flails to agitate the materials, break up clumps of particles, and maintain porosity. Some machines include a conveyor to move the compost. The machines work automatically without an operator and are controlled with limit switches.

Most commercial systems include a set of aeration pipes or an aeration plenum recessed in the floor of the bed and covered with a screen and/or gravel. Between turnings, aeration is supplied by blowers to aerate and cool the composting materials. Since the materials along the length of the bed are at different stages of composting, the bed is divided into different aeration zones along its length. Several blowers are used for one bed. Each blower supplies air to one zone of a bed and is individually controlled by a temperature sensor or time clock.

The capacity of the system is dependent on the number and the dimensions of the beds. The width of the beds in commercially available systems ranges from 6 to 20 feet, and bed depths are between 3 and 10 feet. The beds must conform to the size of the turning machine, and the walls must be especially straight. To protect equipment and control composting conditions, the beds are housed in a building or a greenhouse or, in warm climates, just covered by a roof.

The length of a bed and frequency of turning determine the composting period. If the machine moves the materials 10 feet at each turning and the bed is 100 feet long, the composting period is ten days with daily turning. It increases to twenty days if turning occurs every other day. Suggested composting periods for commercial agitated bed systems range from two to four weeks, though a long curing period may be necessary.

Silos: Another in-vessel technique resembles a bottom-unloading silo. Each day an auger removes composted material from the bottom of the silo and a mixture of raw materials is loaded at the top. The aeration system blows air up from the base of the silo through the composting materials. The exhaust air can be collected at the top of the silo for odour treatment.

A typical composting time for this method might be fourteen days, so onefourteenth of the silo volume must be removed and replaced daily. After leaving the silo, the compost is cured, often in a second aerated silo. This system minimizes the area needed for composting because the materials are stacked vertically. However, the stacking also presents compaction, temperature control, and airflow challenges which must be overcome. Because materials receive little mixing in the vessel, raw materials must be well mixed when loaded into the silo.

Rotating Drums: A different system uses a horizontal rotary drum to mix, aerate, and move the material through the system. The drum is mounted on large bearings and turned through a bull gear. A drum 11 feet in diameter and 120 feet long has a daily capacity of approximately 50 tons with a residence time of three days.

In the drum, the composting process starts quickly; and the highly degradable, oxygen-demanding materials are decomposed. Further decomposition of the material is necessary and is accomplished through a second stage of composting, usually in windrows or aerated static piles.

In some commercial systems, the composting materials spend less than one day in the drum. In this case, the drum primarily serves as a mixing device.

Air is supplied through the discharge end and is incorporated into the material as it tumbles. The air moves in the opposite direction as the material. The compost near the discharge is cooled by the fresh air. In the middle, it receives the warmed air, which encourages the process; and the newly loaded material receives the warmest air to initiate the process.

The drum can be either open or partitioned. An open drum moves all the material through continuously in the same sequence as it entered. The speed of rotation of the drum and the inclination of the axis of rotation determine the residence time. A partitioned drum can be used to manage the composting process more closely than the open drum. The drum is divided into two or three chambers by partitions. Each partition contains a transfer box equipped with an operable transfer door.

At the end of each day's operation, the transfer door at the discharge end of the drum is opened and the compartment emptied. The other compartments are then opened and transferred in sequence, and finally a new batch is introduced into the first compartment. A sill in place at each of the transfer doors retains 15% of the previous charge to act as an inoculum for the succeeding batch. Upon discharge, the compost can go directly into a screen to remove oversized particles which can be returned to the drum for further composting.

Transportable Containers: A different type of in-vessel system, relies on a transportable vessel and a central composting facility. A number of local farms participate and provide manure as a raw material. Each farm receives a transportable vessel, which resembles a solid waste roll-off container. The container has aeration pipes in its base, which are connected to a blower. At the farm, the manure and dry amendments are loaded daily into the container and aerated for several days until the container is picked up and delivered to the central facility to finish composting. When the composting container is picked up, the farm is provided with another empty container to continue the cycle. The farm supplies the manure and receives bulking agent, compost, and/or revenue in return.

Worms

Vermicomposting method

The term "vermicomposting" had recently been coined to mean the use of earthworms for composting organic residues. Earthworms can consume practically all kinds of organic matter and they can eat their own body weight per day; thus, for example, one kilogram of worms can consume one kilogram of residues every day. The excreta or "castings" of the worms are rich in nitrate, available forms of phosphorus, potassium, calcium and magnesium. The passage of soil through earthworms promotes bacterial and actinomycetes growth; actinomycetes thrive well in the presence of worms and their content in worm casts is over six times more than in the original soil.

Kind of worms: A moist compost heap of 2.4 m by 1.2 m and 0.6 m high can support a population of more than 50 000 worms. The introduction of worms into a compost heap has been found to mix the materials, aerate the heap and hasten decomposition. Turning the heaps is not necessary if earthworms are present to do the mixing and aeration. The ideal environment for the worms is a shallow pit and the right sort of worms are necessary. Lumbricus rubellus (the red worm) and Eisenia foetida are thermo-tolerant and so particularly useful. Field worms

Allolobophora caliginosa and night crawlers (Lumbricus terrestris) will attack organic matter from below but the latter do not thrive during active composting, being killed more easily than the others at high temperature.

Vermicomposting in Philippines: The worms used are Lumbricus rubellus and/or Perionyx excavator. The worms are reared and multiplied from a commercially-obtained breeder stock in shallow wooden boxes stored in a shed. The boxes are approximately 45 cm x 60 cm x 20 cm and have drainage holes; they are stored on shelves in rows and tiers.

A bedding material is compounded from miscellaneous organic residues such as sawdust, cereal straw, rice husks, bagasse, cardboard and so on, and is moistened well with water. The wet mixture is stored for about one month, being covered with a damp sack to minimize evaporation, and is thoroughly mixed several times. When fermentation is complete, chicken manure and green matter such as ipil ipil leaves or water hyacinth is added.

The material is placed in the boxes and should be sufficiently loose for the worms to burrow and should be able to retain moisture. The proportions of the different materials will vary according to the nature of the material but a final protein content of about 15% should be aimed at. A pH value as near neutral as possible is necessary and the boxes should be kept at temperatures between 20C and 27C. At higher temperatures, the worms will aestivate and, at lower temperatures, they hibernate.

In spite of their being able to eat the bedding material, the worms at this stage are fed regularly; for every kilogram of worms a kilogram of feed is given every 24 hours. For each 0.1 mof surface area, 100 g of breeder worms are added to the boxes. The feedstuffs used are again various and include chicken manure, ipil ipil, vegetable wastes and so on. At one farm, water hyacinth is grown specifically and used fresh (chopped up) as the sole source of feed.

Composting procedure:

— A series of pits (the number depending on available space) are dug approximately 3 m x 4 m x 1 m deep, with sloping sides. Bamboo poles are laid in a parallel row on the pit floor and covered with a lattice of wood strips. This provides the necessary drainage as the worms cannot survive in a waterlogged environment.

— The pit is then lined with a suitable material to keep the worms from escaping into the surrounding soil (although, with the abundant feed provided in the compost heap, this may not happen) and yet permit drainage of excess water. At the farm under consideration, old animal feedstuff sacks were used.

— The pit can now be filled with rural organic residues such as straw and other crop residues, animal manure, green weeds, leaves and so on. The filled pit is covered loosely with soil and kept moist for a week or so. During this period, another pit can be filled as necessary.

— One or two spots on the heap are then well watered and worms from the breeding boxes are place on top; the worms immediately burrow down into the damp soil.

— To harvest the worms from the boxes, two-thirds of the box is emptied into a new box lined with banana leaf or old newspaper. The original box can now be provided with fresh bedding material and those worms remaining will again multiply. The worms emptied from the box are picked out by hand for adding to the heap.

— The compost pit is left for a period of two months; ideally it should be shaded from hot sunshine and it must be kept moist. Within two months, about 10 kg of castings will have been produced per kilogramme of worms. The pit is then excavated to an extent of about two-thirds to three-quarters and the bulk of the worms removed—by hand or by sieving. This leaves sufficient worms in the pit for further composting and the pit can be refilled with fresh organic residues. The compost can be sun-dried and sieved to give a very good quality material. A typical analysis is: Organic matter 9.3%, Nitrogen 8.3%, Phosphorus 4.5%, Potassium 1.0% (water-soluble), Calcium 0.4%, Magnesium 0.1%.

— The excess worms that have been harvested from the pit can be used in other pits, sold to other farmers for the same purpose, used or sold for use as animal feed supplement, used or sold for use as fish food or, if there is no social taboo, used in certain human food preparations.

Vermicomposting in Cuba: In Cuba, different methods are used for worm propagation and vermicomposting.

Worm troughs in a row: The first and most common is cement troughs, two feet wide and six feet long, much like livestock watering troughs, used to raise worms and create worm compost. Because of the climate, they are watered by hand every day. In these beds, the only feedstock for the worms is manure, which is aged for about one week before being added to the trough. First, a layer of three to four inches of manure is placed in the empty trough, then worms are added. As the worms consume the manure, more manure is layered on top, roughly every ten days, until the worm compost reaches within a couple inches of the top of the trough, about two months. Then the worms are separated from the compost and transferred to another trough.

Windrows: The second method of vermicomposting is windrows. Cow manure is piled about three feet across and three feet wide. Then it is seeded with worms. As the worms work their way through it, fresh manure is added to the end of the row, and the worms move forward. The rows are covered with fronds or palm leaves to keep them shaded and cool. Some of these rows have a drip system—a hose running alongside the row with holes in it. But mostly, the rows are watered by hand. Some of these rows are hundreds of feet long. The compost is gathered from the opposite end when the worms have moved forward. Then it is bagged and sold. Fresh manure, seeded with worms, begins the row and the process again.

Some of the windrows have bricks running along their sides, but most are simply piles of manure without sides or protection. Manure is static composted for 30 days, then transferred to rows for worms to be added. After 90 days, the piles reach three feet high. Worm populations, they say, can double in 60 to 90 days. Windrows are also used to compost rice hulls and sugar cake (cake is what is left after sugar cane is processed), but this too is mixed with animal manure. Sometimes food scraps added to worm beds.

Vermiculture in India:

Preparing vermicompost

— Materials-breeder worms, a wooden bed and organic wastes.

— The bed should be 2 ½ ft. high x 4 ft. wide x any length desired. Apply worms for every part of waste.

— Sieving and shredding-Decomposition can be accelerated by shredding raw materials into small pieces.

— Blending- Carbonaceous substances like sawdust, paper and straw can be mixed with nitrogen rich materials such as sewage sludge, biogas slurry and fish scraps to obtain a near optimum C/N ratio of 30:1 / 40:1. A varied mixture of substances produces good quality compost, rich in major and micro nutrients.

— Half digestion- The raw materials should be kept in piles and the temperature allowed to reach 50-55C. The piles should remain at this temperature for 7 to 10 days.

— Moisture, temperature and pH- The optimum moisture level for maintaining aerobic conditions is 40-45%. Proper moisture and aeration can be maintained by mixing fibrous with nitrogen rich materials. The temperature of the piles should be within 28-30C. Higher or lower temperatures will reduce the activity of micro flora and earthworms. The height of the bed can help control the rise in temperature. The pH of the raw material should not exceed 6.5 to 7.

After about a month the compost is ready. It will be black, granular, lightweight and humusrich. To facilitate separating the worms from the compost, stop watering two to three days before emptying the beds. This will force about 80% of the worms to the bottom of the bed. The rest of the worms can be removed by hand. The vermicompost is then ready for application.

Bibliography

Anderson, Robert G. and Robert Hadad, "Nutrient analysis of organic fertilizers for greenhouse vegetable production", *HortScience*, 1999.

Anon, "Colored mulch improves yields", *Greenhouse Management & Production*, 1995.

Batal, K.M., et al., "Effects of winter cover crops and N applications on vegetable crop production systems", *HortScience*, 1995.

Byczynski, Lynn., "Greenhouse tomato considerations", *Growing for Market*, 1993.

Creamer, N.G., B. Plassman, M.A. Bennett, R.K. Wood, B.R. Stinner, and J. Cardina, *A Method For Mechanically Killing Cover Crops To Optimize Weed Suppression*, Ohio Agriculture Research and Development Center, Ohio State University, Wooster, OH. In press, 1995.

Csizinszky, A.A. et al., "Evaluation of color mulches and oil sprays for yield and silverleaf whitefly control on tomatoes", *HortScience*, 1995.

Grossman, Joel, "Sweetcorn in tomatoes keeps the fruitworm at bay", *Organic Gardening*, 1980.

Hazzard, Ruth and Robert Wick, "Management of early blight & septoria leaf spot in fresh market tomatoes", *The Natural Farmer*, 1996.

Lumsden, Robert D. et al., "Effect of organic amendments on soilborne plant diseases and pathogen antagonists", In: Lockeretz, William (ed.) *Environmentally Sound Agriculture (Selected Papers from the Fourth International Conference of IFOAM)*, Praeger Press, New York, NY, 1983.

McMullin, Eric and Warren Clark, "Riding Israel's vegetable boom", *American Vegetable Grower*, 1993.

Morgan, Lynette, "Organic fertilizers for hydroponics", *The Growing Edge*, 1997.

Rosen, Emily Brown, "Dilution Rates for Hydrogen Peroxide", *Northeast Organic Farmers Association*-New Jersey, 1995.

Schonbeck, Mark, "Mulching choices for warm-season vegetables", *The Virginia Biological Farmer*, 1995.

Schwankl, Lawrence J., and Glenn McGourty, "Organic fertilizers can be injected through low-volume irrigation systems", *California Agriculture*, 1992.

Tourte, Laura, "Kelp extract and fish powder sprays on organically grown processing tomatoes", *Organic Farming Research Foundation Information Bulletin*, 1997.

Weltzein, Heinrich C., "Some effects of composted organic materials on plant health", *Agriculture, Ecosystems and Environment*, 1989.

Index